Henry Reid

The Science and Art of the Manufacture of Portland Cement

With Observations on some of its constructive Applications

Henry Reid

The Science and Art of the Manufacture of Portland Cement
With Observations on some of its constructive Applications

ISBN/EAN: 9783337106461

Printed in Europe, USA, Canada, Australia, Japan

Cover: Foto ©ninafisch / pixelio.de

More available books at **www.hansebooks.com**

THE SCIENCE AND ART

OF THE MANUFACTURE OF

PORTLAND CEMENT.

WITH

OBSERVATIONS ON SOME OF ITS CONSTRUCTIVE
APPLICATIONS.

BY

HENRY REID, C.E.,

AUTHOR OF
'A PRACTICAL TREATISE ON THE MANUFACTURE OF PORTLAND CEMENT,'
AND 'A PRACTICAL TREATISE ON CONCRETE MAKING.'

LONDON:
E. & F. N. SPON, 46, CHARING CROSS.

NEW YORK:
446, BROOME STREET.

1877.

THIS BOOK

IS

𝔇𝔢𝔡𝔦𝔠𝔞𝔱𝔢𝔡 𝔱𝔬 𝔱𝔥𝔢 𝔐𝔢𝔪𝔬𝔯𝔶

OF

JOHN SMEATON,

THE

FAMOUS CONSTRUCTOR OF

THE EDDYSTONE LIGHTHOUSE.

PREFACE.

DURING the last ten years great progress has been made in the knowledge of Portland Cement, and its use has in consequence been extended in a remarkable degree. Engineering works, more especially in the direction of docks and harbours, have given an impetus to its application, securing for it a position of unchallenged importance. This increasing appreciation of its valuable constructive properties has secured a more reliable quality, which the manufacturers are now ambitious to maintain. Since 1868, when my book on the manufacture of Portland Cement was published, the production of Cement has more than doubled in the London district alone.

This gratifying increase, and the accompanying prosperity of the Cement-maker, has not yet led to much improvement in the processes of manufacture, so as to cheapen its cost or render it more acceptable to the consumer. In Germany a more healthy feeling prevails, and no effort is wanting on the part of the manufacturers of Portland Cement to extend the range of its usefulness by the adoption of the best and latest improvements which technical skill can devise. In furtherance of this object the mutual co-operation of makers and users is secured with such profitable results as a confiding interchange of ideas must command. The first fruit of such a combination has resulted in the establishment of uniform testing on a basis sound enough to secure accurate returns, controlled by the best available scientific guidance. When by the process of evolution or some other force of "natural development" such a position is reached in England, all concerned will then be able to repose in a haven of comfort, sheltered from the storm of amateur experimenters and their bewildering creations.

I have in the following pages, from the best available sources, discussed the question of Portland Cement-making in its numerous and varied bearings. Its geological aspect has been freely noticed, with the object of removing the erroneous impression or belief that Portland Cement is only obtainable from a combination of chalk and river clays. The prominence given to that division of my argument will, I hope, direct serious attention to the easy production of Cement from sources hitherto regarded as quite incapable of furnishing the necessary materials for its successful fabrication. The treatment of these more obdurate materials naturally involves the use of machines differing widely in character from the wash-mill of the "chalk and clay" process. The machinery of reduction required for this purpose has therefore been noticed at some length, and the various machines and their characteristics reviewed and illustrated.

I have discussed the question of Cement-burning in its purely scientific direction, and compared some of the more familiar methods, with an examination of the various kilns, and their leading features and character.

Cement-testing, and the machines used for that purpose, have been referred to at considerable length, but not, in my opinion, more fully than the importance of the subject requires. If the observations on, and review, of the Cement experiments should seem to some of my readers hypercritical, I would plead that being in a great measure responsible for the original adoption of the "Board of Works'" test, I am anxious, if possible, to arrest its reaching a point where reason or common sense cannot follow. If this tendency of aiming at the ideal and impracticable continues unchecked, the eventual result will be a refusal to be controlled by an ever restless and changing system, the ultimate limits of which admit of no reasonable estimation.

Although noticing generally some of the numerous applications of Concrete, I have more particularly referred to the latest outcome of this material in its useful adaptation to paving and sewering. The progress already and so speedily made in this direction, in which all are so intimately concerned, indicates that even now we are still far distant from the ultimate usefulness of which it is capable. In this work, which treats more directly of the Cement, on the quality of which so much of the excellence of Concrete depends, it is impossible to notice at great length the numerous purposes to which it is applied. However, I intend shortly to follow up the subject in a work specially devoted to the consideration of Concrete construction in all its varied and interesting developments.

In a work of this character it would be mere affectation to pretend that its matter and details are altogether original. On the contrary, I am much indebted to many authorities for analyses and other information of the most valuable kind. When such extracts have been used I have, where it was necessary, acknowledged the source of their origin, not so much from the proper desire to record my indebtedness, as to add value to a work which, notwithstanding my earnest endeavours, cannot but fail to be regarded as imperfect in many of its parts.

HENRY REID.

LONDON, *October*, 1877.

CONTENTS.

CHAPTER I.
HISTORICAL AND INTRODUCTORY .. ..   PAGE 1

CHAPTER II.
GEOLOGICAL AND MINERALOGICAL OBSERVATIONS 31

CHAPTER III.
ARTIFICIAL SOURCES FROM WHICH SOME CEMENT-MAKING MATERIALS MAY BE OBTAINED 120

CHAPTER IV.
CONSIDERATIONS WHICH SHOULD INFLUENCE THE MANUFACTURER IN SELECTING A SITE FOR THE CEMENT WORKS 135

CHAPTER V.
OBSERVATIONS ON THE SELECTION OF THE RAW MATERIALS 140

CHAPTER VI.
THE ESTIMATION OF THE RAW MATERIALS 148

CHAPTER VII.
TREATMENT OF THE RAW MATERIALS 157

CHAPTER VIII.
THE MANUFACTURE OF PORTLAND CEMENT FROM CHALK AND CLAY .. 162

CHAPTER IX.
THE MANUFACTURE OF PORTLAND CEMENT FROM THE BLUE LIAS MATERIALS 168

CHAPTER X.

THE MANUFACTURE OF PORTLAND CEMENT FROM THE CARBONIFEROUS AND OTHER LIMESTONES 178

CHAPTER XI.

MACHINERY OF REDUCTION 182

CHAPTER XII.

THE WASH-MILL 207

CHAPTER XIII.

MILLSTONES 213

CHAPTER XIV.

BACKS OR RESERVOIRS 229

CHAPTER XV.

DRYING OVENS 234

CHAPTER XVI.

KILNS AND MODE OF BURNING 239

CHAPTER XVII.

BRICK-FORMING MACHINERY 262

CHAPTER XVIII.

TESTING MACHINERY 274

CHAPTER XIX.

CEMENT TESTING 294

CHAPTER XX.

CARELESS USE OF CEMENT 350

CHAPTER XXI.

THE VARIOUS USES TO WHICH PORTLAND CEMENT IS APPLIED 362

INDEX 410

LIST OF ILLUSTRATIONS.

Portrait of Author		*Frontispiece.*
Portrait of Smeaton		PAGE xiii
Fig. 1.—Schöne's Clay-tester		149
,, 2.—Dr. Scheibler's Carbonic Acid Tester		154
,, 3.—Original "Blake" Stone-breaker		184
,, 4.—Marsden's Improved "Blake"		187
,, 5.—Hall's ,, ,,		189
,, 6.— ,, ,, ,,		190
,, 7.— ,, ,, ,,		190
,, 8.—Broadbent's ,, ,,		193
,, 9.—Gray's Excelsior Stone-breaker		194
,, 10.— ,, ,, ,,		195
,, 11.— ,, ,, ,,		196
,, 12.—Archer's Stone-breaker and Pulverizer		197
,, 13.—Goodman's Crusher and Triturator		199
,, 14.— ,, Double-action Crusher		203
,, 15.—Improved Single-action Crusher		204
,, 16.—Goodman's New Double-action Crusher		205
,, 17.—Harrow Wash Mill		208
,, 18.—Knife ,, ,,		209
,, 19.—Millstone Building		215
,, 20.—Portable Grinding Mill		220
,, 21.—Straub and Co.'s Scientific Grinder		224
,, 22.—Common Dome Kiln		240
,, 23.—Hoffman's Kiln		247
,, 24.— ,, ,,		247
,, 25.—Bock's ,,		256
,, 26.— ,, ,,		256
,, 27.— ,, ,,		256

LIST OF ILLUSTRATIONS.

		PAGE
,,	28.—Brick-forming Machine	263
,,	29.—Fairburn's Steam Brick-press	266
,,	30.—Guthrie's Spring ,,	268
,,	31.— ,, ,, ,,	270
,,	32.— ,, ,, ,,	270
Figs. 33 and 35.—Vicat's Prism Tester		276
Fig. 34.—Vicat's Needle Test		276
,,	36.— ,, Improved Needle Test	277
,,	37.—Pasley and Treussart Tester	278
,,	38.—Adie's Testing Machine	279
,,	39.—Pallant's Testing Machine	280
,,	40.—Michele's ,, ,,	281
,,	41.—Thurston's ,, ,,	282
Figs. 42 and 43.—Fruhling's Tensile Testing Machine		284
Fig. 44.—Fruhling's Compression Tester		285
,,	45.—Bailey's Tensile Tester	286
,,	46.— ,, Hydraulic Tester	287
,,	47.—Stettin Testing Machine	288
,,	48.—Leger and Aron's Testing Machine	291
,,	49.—Briquette Press	297
Figs. 50 and 51.—Mann's Gravimeter		299
Fig. 52.—Nicholsons Portable Balance		305
,,	53.—Deacon's proposed Tester	317
,,	54.—Diagram of Twelve Months' Test	336
,,	55.— ,, Ten Years' Test	338
,,	56.—View of Wrecked Cottages	353
,,	57.—Washing Granite Gravel	383
,,	58.—Moulding Concrete Slabs	384
,,	59.—Silica Tanks	385
,,	60.—View of Victoria Stone Works	386
,,	61.— ,, Hodges and Butler's Works	397
Figs. 62 and 63.—Pipe-making Machine		398
Fig. 64.—Views of Silicated Stone Pipes		400
,,	65.— ,, Pipes under Test	401
,,	66.— ,, Pipes and Slabs	405

JOHN SMEATON,
CIVIL ENGINEER.
Born 1724; died 1792.

EDDYSTONE LIGHTHOUSE.

EXPLANATORY.

RECENTLY and some time after this book had gone to press, Mr. Douglas at the meeting of the British Association publicly disclosed the intention of the Trinity Board to dismantle the famous Eddystone Lighthouse. So startling a proposition has naturally created in the public mind a feeling of surprise, which has, however, been somewhat allayed by the statement that the fabric itself has not succumbed to the elemental strife with which it has so long contended. The rock on which it stands is no longer stable enough to maintain its famous companion, and that circumstance, together with the increasing demands of commerce, necessitates a building of greater extent, capable of receiving the most improved danger-warning apparatus for the safe guidance of our own and other ships of any and every nationality.

Having dedicated this book to the memory of the illustrious engineer by whom this world-famous building was erected, we consider it proper to insert here an article from the "Daily News" of the 23rd August, which fully describes the nature and necessity of this step on the part of the Trinity Light Board.

"'After having been buffeted by the storms of eighty years'— such are the words of an enthusiastic tribute to the work of a great engineer, which was then but eighty years in existence— 'the Eddystone stands unmoved as the rock it is built on—a proud monument to its great author.' There is something significant in the words which declare the Eddystone Lighthouse to be unmoved as the rock it was built on. The allusion to John Smeaton and his great work appears in the first volume of the Transactions of the Institution of Civil Engineers. The recent proceedings of the Mechanical Science Department of the British Association inform us that the Lighthouse is to be taken down because of the undermining of the rock on which it stands. It has thus proved itself unmoved as the rock; but when the rock is undermined, the Eddystone Lighthouse must submit to the necessity of change. A paper was read to the section by Mr. J. N. Douglas, Engineer to the Trinity House, explaining the facts which make it needful to take down Smeaton's lighthouse and erect a new one; and the president of the section, Mr. Edward Woods, expressed the common regret and concern that 'the glory of Smeaton' is about to become one of the things of the past. 'They had all hoped and believed,' he said, 'that the Eddystone Lighthouse would endure throughout all time;' but, instead, it is only to be another illustration of the fact that men do not build for ever. It was not in any spirit of extravagant rhetoric that Smeaton's lighthouse was called his glory. It was indeed one of the glories of English engineering science. Our necessities, and with them our inventions, have grown, and it seems that the elevation and the range of lights which Smeaton's tower allows, are not adequate either to the wants or to the scientific capabilities of our time. When the new lighthouse is erected, a first-class fog-signal will, it seems, probably be introduced, for which the dimensions of Smeaton's tower are not adequate. Mr. Douglas expressed a hope that when the famous tower is taken down it may be put up again somewhere on the shore as a national monument. Certainly few buildings erected in England during many centuries could better deserve to be regarded as a national monument in the truest sense. Smeaton's work is a monument of English engineering genius, invention, and success. The volume from which we have already quoted observes that buildings of the same kind as Smeaton's lighthouse have been erected since—and indeed our ideas of engineering have considerably expanded since even those words were written—'but it should always be borne in mind who taught the first great lesson, and recorded the progressive steps with a modesty and simplicity that may well be held up as models.' The tribute to the modesty and simplicity of Smeaton's personal character is well deserved. In dedicating his own account of the building of the lighthouse to George III., he observes:—'I can with truth say I have ever been employed in works tending to the immediate benefit of your Majesty's subjects.' This was the simple truth. Little of his personal history is known, but enough is made public to show that he was unassuming and unselfish, and that he did his great work for good, and not for either fame or gain.

"There used to be a story popular among the children, who are elderly men and women now, about a group of young people sitting round a fire and talking of their favourite heroes. One admired Julius Cæsar; another Alexander the Great; another, of a less warlike spirit, Christopher Columbus; and so on, until the turn came to one little philosopher, who declared that they who liked might have their conquerors, but for him his hero was John Smeaton. Naturally the others of the little group had never heard of John Smeaton, and doubtless much of the effect of the story would have been lost if the readers had not proved to be in a similar state of ignorance. But the young philosopher went on to enlighten his hearers; told them of John Smeaton's great work, and what good it had done for mankind; and, we need hardly say, succeeded in the end in convincing all his listeners that Smeaton's was a success which properly ought to throw into the shade the triumphs of heroes and kings. We are not concerned to discuss the value of the familiar moral pointed thus by the youthful admirer of Smeaton; but assuredly to those who hold that heroism is only to be tested by the amount of practical good it does, and is the first to do, the name of the builder of the Eddystone Lighthouse might serve as a very suitable illustration of their doctrine. It would not be easy to mention any great conqueror whose triumphs cost so many lives as the work of Smeaton may have saved. The Eddystone Lighthouse which he built was not the first, nor even the second, which stood upon the rock that gave it a name. The first lighthouse put there was that begun under Mr. Winstanley in 1696, which was finished in three years, and lasted but four years more. A winter tempest in 1703 destroyed it, and its builder perished with it. A second structure, which was formed of wood, for the sake of lightness, was put up a few years after. Fire destroyed this in 1755: and then the genius of Smeaton discovered the way by which a tower could be perched upon the rock which could hold its own against the winds and waves, and yet defy fire as well. Even Smeaton's own structure was found afterwards to have too much woodwork about it, and a great part of its construction had to be modified by him in consequence of an imperious hint from the unsparing master which had destroyed the second edifice. But the great thing that Smeaton had done was to show how a lighthouse could be so constructed as to stand on a lonely

b

rock in the midst of fierce waves, and to hold its erect position there while the furious winds, breaking all around, were shivering the spars of ships and rending branches from trees that stood far inland on the shore. A lighthouse to be of any use in such a place must be tall and stately. It must stand unsheltered, so as to send its light streaming far across the midnight sea. It must be enduring as the rock, and yet it must not present the rock's broad front. It must be at once slender and capacious. It must offer the least possible resistance to the storm, and yet be strong enough to stand up in any gale. These are the various and sometimes seemingly incompatible properties needed for a lighthouse such as the Eddystone: and it is due to the genius and the courage of Smeaton that England first knew how to combine them with something like a certainty of efficiency and success.

"Smeaton accomplished other great works as well as the building of the lighthouse, but his fame will always rest on that great achievement. He was, as an engineer, somewhat cast in the mould from which George Stephenson came forth. He was equal to any risk indeed, but he was essentially practical, and the success he looked to was that of really solid, valuable work. He did not care for merely brilliant enterprises. The romance of the profession, that has proved so tempting and often so fatal to the genius of great engineers, had little charm for him. He was deliberate and cautious in all he did—that is, until he thoroughly saw his way—and then he was bold enough for anyone. In engineering science as in war, there seem two great classes of leaders; the brilliant and adventurous, and the steady and safe : the Napoleon and the Wellington. Smeaton, like Stephenson, belonged to the latter class. There is something of the wild and the poetic in the very idea of his lighthouse perched upon its rock in the midst of the stormy waters. It might stand as the emblem of daring and of defiance. But it was the result of the most careful thought and cautious examination, and its builder scarcely took one step forward without having made the success of the movement certain. Its endurance justified in every way the anticipation and the confidence of its author. It yields now, if it is to yield, only to the operation of causes which affect in no manner his calculations. He established a principle which survives its particular application. He had a positive genius for his work.

Those who tell us of his life, say that when he was a child his toys were the tools of men. Like all manner of poets, philosophers, and artists he had tried hard to be a lawyer and failed; his mind refused to 'bite into' any craft but that for which his very infancy had indicated such a capacity. His fame was like his work, solid and staunch. A great writer has compared men of a certain brilliant and superficial class to the lighthouse, which he says, is over high, far-shining, and empty. But it is only by the utmost stretch of metaphor that the lantern of the lighthouse can be considered as empty, full as it is of the light that is the guidance and safety of unending generations of voyagers. Assuredly no fame and no labours could be less fittingly illustrated by anything that symbolises emptiness than those of Smeaton. His name was popular everywhere in England. Thousands of Englishmen who had never seen the Eddystone Lighthouse were in the habit of citing it as a type of solidity, endurance, and success. The new building will no doubt be a great improvement on the old one, for the builder will come to his task with the advantage of all that engineering science has added since Smeaton's day. But whatever the structure that supplies the place of the 'Eddystone,' it might as well bear carved on it the name of Smeaton; for that name will always be associated with the spot and be preserved by the genius of the place."

THE SCIENCE AND ART
OF
PORTLAND CEMENT.

CHAPTER I.

HISTORICAL AND INTRODUCTORY.

THERE can be no doubt that the experiments of Smeaton in 1757 lightened up the darkness surrounding the subject of mortars, and their behaviour under varied circumstances. The necessity for a truly reliable mortar for the building of the Eddystone Lighthouse, capable of withstanding the influence of sea water, and possessing also the capacity of resisting the mechanical action of violent storms, was a problem of much difficulty, but which the illustrious engineer successfully solved. It is needless for us to recite the dangers surrounding the erection of a structure on the site of which the previous erections had been destroyed, and the successor of which still remains to testify, for more than a century, to the ability and foresight displayed in its conception and execution. Those who have read the story of that work during its preliminary stages, and the protracted and dangerous process of construction, will readily understand the anxiety of the directing engineer, on whom the whole responsibility devolved. Our allusion to it will only have reference to the mortar experiments for the purpose of showing the careful investigation and thoughtful reasoning displayed in the selection of the mortar, and as an example for others under similar circumstances to follow. There is just as much necessity now for the engineer to adopt the same precautions as then, although we fear that those now representing the engineering profession would regard such a process of examination beneath their consideration. It is just possible that many in this country may fail to appreciate the great value of those experiments, but we are glad to find that elsewhere very recent testimony of a most laudatory character fully testifies to the value of Smeaton's exertions. We avail ourselves of this opportunity of putting it on record.

In a work on 'Hydraulic Mortars,' &c. (published at Leipsig in 1869, by Dr. Michaelis), the following passage occurs:

"A century has elapsed since the celebrated Smeaton completed the building of the Eddystone Lighthouse. Not only to sailors, but to the whole human race is this lighthouse a token of useful work, a light in a dark night. In a scientific point of view, it has illuminated the darkness of almost two thousand years.

"The errors which descended to us from the Romans, and which were even made by such an excellent author as Belidor, were dispersed.

"The Eddystone Lighthouse is the foundation upon which our knowledge of hydraulic mortars has been erected, and it is the chief pillar of modern architecture. Smeaton freed us from the fetters of tradition by showing us that the purest and hardest limestone is not the best, at least for hydraulic purposes, and that the cause of hydraulicity must be sought for in the argillaceous admixture.

"It was a long time before men of science confirmed this statement of the English engineer, or corrected the ideas on the hardening of hydraulic mortars, which were then necessarily confused on account of the imperfect state of chemistry at that time. How could science subsequently keep pace with practical progress? for even at present, though we have possessed for about half a century the most excellent hydraulic mortars, the hardening process is not yet completely explained."

Pasley also, in the preface to the first edition of his work, dated 17th September, 1838, says of Smeaton, in very similar language, as follows:

"Of all the authors who have investigated the properties of calcareous mortars and cements, from time immemorial to the present day, our countryman, Smeaton, appears to me to have the greatest merit; for although he found out no new cement himself, he was the first who discovered, in or soon after the year 1756, that the real cause of the water-acting properties of limes and cements consisted in a combination of clay with the carbonate of lime; in consequence of having ascertained by a very simple sort of chemical analysis that there was a proportion of the former ingredient in all the natural limestones, which, on being calcined, developed that highly important quality, without which walls exposed to water go to pieces, and those exposed to air and weather only are comparatively of inferior strength. By this memorable discovery Smeaton overset the prejudices of more than two thousand years, adopted by all former writers, from Vitruvius in ancient Rome to Belidor in France and Semple in this country, who agreed in maintaining that the superiority of lime consisted in the hardness and whiteness of the stone, the former of which may or may not be accompanied by water-setting or powerful cementing properties, and the latter of which is absolutely incompatible with them. The new principle laid down by Smeaton, the truth of which has recently been admitted by the most enlightened chemists and engineers of Europe, was the basis of the attempts made by Dr. John, at Berlin, and by M. Vicat (the engineer) in France, to form an artificial water-lime or hydraulic lime, in 1818, and of mine to form an artificial water-cement at

Chatham, in 1826, to which I was led by the persual of Smeaton's observations, without knowing anything of the previous labours of these gentlemen on the Continent, or of Mr. Frost, the acknowledged imitator of M. Vicat, in this country."

There can be no question of the indebtedness the science of water cements is under to Smeaton, and the above-quoted writers have unhesitatingly acknowledged what is now universally regarded as Smeaton's well-deserved due. A more than perhaps necessary reference to this subject may tend to stimulate engineers and chemists to a still further examination of cements and their peculiarities, and may thus develop some new properties of which we are still ignorant. Silica and, in a minor degree, alumina, which now play only a subordinate part (as far as volume is concerned) in connection with carbonate of lime, may be found to be capable of occupying a more important position in the fabrication of hydraulic mortars and cements.

It is pleasing to find that the engineer and the lighthouse are so closely identified, and no one cares to recollect under what reign it was built. The name of the engineer and not that of the king descends to posterity, receiving the credit of his labours. Sostratus, the architect who built the famous tower of Pharos, near Alexandria, on the top of which a fire was kept burning to guide such ships as sailed at night in those dangerous coasts, resorted to a stratagem to hand down to posterity the fact of its being his handiwork. The potentate (Ptolemy Philadelphus), under whose authority and command the tower was erected, had the usual inscription placed thereon, but it was very evanescent in character, being cut on a coating of lime, under which was carved in the marble the following dedication :

"Sostratus, the Cnidian, son of Dexiphanes, to the protecting deities, for the use of seafaring people."

A modern historian, however, did not appreciate this subterfuge, and thus remarks :

"The lime soon mouldered away; and by that means, instead of procuring the architect the honour with which he had flattered himself, served only to discover to future ages his mean fraud and ridiculous vanity."

A long interval of time has elapsed between the building of the first lighthouse and that which is so intimately identified with Smeaton's name.

It would be tedious to follow, in Smeaton's own words, the process by which he succeeded in eliminating from the then prevailing dogmatism the necessary data, even for a base or starting point on which to build his initiatory experiments. Up to this time varied proportions of lime and tarras were used in the preparation of mortars for engineering works in water. The practical masons of those days were of opinion that the harder the stone from which the lime was made, the better would it set or harden. It was also a common impression that if the mortar was made with sea water it would never harden in so great a degree as that made from fresh water. Smeaton's own observation had informed him that where plastering in house building was done by using sea water such work never became thoroughly dry, and was influenced by atmospheric change in a most prejudicial manner. The same injurious action occurred when sea sand was used in the mortar. Corresponding influences are still at work under like conditions of improper construction. There was no fresh water on the Eddystone rock, and only a comparatively pure limestone on the mainland, so that he directed his attention to the experiments with limes obtained from a distance. But we had better, before proceeding farther, give his own description of the mode he adopted in preparing for the experiments.

"I took as much of the ingredients as altogether would

ultimately form a ball of about two inches in diameter. This ball, lying upon a plate till it was set and would not yield to the pressure of the fingers, was then put into a flat pot filled with water, so as to be covered by the water; and what happened to the ball in this state was the criterion by which I judged of the validity of the composition for our purposes. The measure I used was a common small chip box, taking as many measures from each ingredient as I meant to try. I constantly put down the lime upon the flat bottom of a common pewter plate, and with as much water as would sufficiently wet it, worked it upon the plate with a broad-pointed knife till it was become a tough, but a pretty soft paste. I afterwards added the quantity I intended of tarras or other gross matter gradually, working it after each addition till it was become tough; and in this way adding the gross matter at three or four different times."

These balls thus prepared were then immersed in water, and his first lot, two parts lime and one part tarras, set in water, although not to his satisfaction. He found at this stage the fallacy of the strongest lime being obtained from the hardest rock, such as the Plymouth marble, and also at this stage convinced himself that these proportions behaved equally well in salt water and fresh. This difficulty therefore was overcome.

Having seen in Wren's 'Parentalia' that cockle or other shell lime had been usefully employed in the construction of St. Paul's Cathedral, he tried it, and found that, without any admixture of tarras or sand it set hard and readily; but on being put into water after it was set, gradually macerated and dissolved. He afterwards learned that this kind of lime had been used in a part of the work of Ramsgate pier, and owing to its solubility in sea water the masonry had to be pulled down.

The Aberthaw lime, from Glamorganshire, with various

proportions of tarras, excelled in a great degree any of the other mixtures, and its behaviour under water was of the most satisfactory character, besides it continued to harden for months. In the early stage of the experiments he had, while trying the hydraulic properties of plaster of Paris, decided that if the mortar used in the building of the lighthouse did not set fast enough he could (and eventually did) point the external and other exposed joints with this quick-setting material, which, although it could not long resist the action of the sea water, would do so sufficiently long to prevent any injurious action on the permanent or mortar joints.

Analysis of Aberthaw Limestone by Phillips.

Carbonate of lime	86·2
Clay	11·2
Water, &c.	2·6
	100·0

The difference in the hardening of the mortars made from the various limestones excited his curiosity, and he became anxious to trace the cause. A friend instructed him how to analyze the limestones, and he thus describes the mode he adopted:

"I took about the quantity of five pennyweights (or a guinea's weight) of the limestone to be tried, bruised to a coarse powder; upon which I poured common *aqua fortis*, but not so much at a time as to occasion the effervescence to overtop the glass vessel in which the limestone was put; and added fresh aqua fortis after the effervescence of the former quantity had ceased, till no further ebullition appeared by any addition of the *acid*. This done, and the whole being left to settle, the liquor will generally acquire a tinge of some transparent colour; and if from the solution little or no sediment drops, it may be accounted a pure limestone (which is generally the case with white chalks and several others), as containing no uncalcareous matter; but if from the solution a quantity of matter is deposited in the form of mud, this indicates a quantity of uncalcareous matter in its composition. When this is well settled pour off the water, and repeatedly add water in the same way, stirring it and letting it settle till it becomes tasteless. After this let the mud be well stirred into the water, and, without giving it time to settle, pour off the muddy water into another vessel; and if there is any sand or gritty matter left behind (as will frequently be the case) this collected by itself will ascertain the quantity and species of fabulous matter that entered into the texture of the limestone. Letting now the muddy liquor settle, and pouring off the water till no more can be got without an admixture of mud, leave the rest to dry; which, when it comes to the consistence of clay, or paste, make it into a ball and dry it for further examination."

The modern chemist would take a more accurate method to ascertain the analysis of limestone, but the simplicity of Smeaton's operation commends itself as a ready means of obtaining an approximate test of unknown limestones, and might even now be profitably performed in the preliminary examination of new and untried districts.

Smeaton proceeds to give the results obtained from his chemical tests, and found chalk and Plymouth marble (as he terms it) were completely dissolved in the acid. Aberthaw limestone, however, resulted in a small deposit of sandy matter, some of which was crystalline in character, and the whole of a dirty appearance. The muddy *residuum* was tough and tenacious, resembling blue clay, which, when made into a ball and burnt, became reddish in colour, equalling a brick in hardness. The total sediment weighed nearly one-eighth of the whole. The Bridstow stone (obtained in Devonshire, about thirty-five miles from Plymouth) when similarly treated gave nearly the same results.

Having thus discovered the cause of hydraulicity in certain limestones, he was satisfied, and accordingly determined to adopt the Aberthaw lime in conjunction with tarras, but he paused, having heard that this material in combination with lime threw out, when constantly under water, concrescences resembling the stalactites found in the caves of the carboniferous limestone formation. A reference to Belidor's 'Hydraulic Architecture' informed him that *Terra Puzzolana* found in Italy near *Civita Vecchia*, made a good water cement. Fortunately he found at Plymouth a considerable quantity of this material, which a merchant had imported on speculation, expecting to sell it to the constructors of old Westminster Bridge. But neither commissioners, engineers, nor contractors would trouble themselves even to try it, so the venture turned out a bad one for the importing merchant, but a lucky thing for Smeaton. His experiments proved it superior to tarras, and in the end he says, "With respect to these balls that were constantly kept under water, they did not seem inclined to undergo any change in form, only to acquire hardness gradually, insomuch that I did not doubt but to make a cement that would equal the best *merchantable Portland stone* in solidity and durability."

It is worthy to note this distinction made by Smeaton. He says *merchantable* Portland stone, implying therefore that all the products of the quarries were not suitable for building purposes. Wren was equally fastidious, and only accepted such stones for his buildings as could be laid on their natural beds. Until recently some of these rejected stones were still to be seen at the quarries.

We have already shown that there can be no question, from Smeaton's own words, that the method he adopted in carrying out his experiments indicated in the clearest manner how it was possible even to convert a limestone into an hydraulic mortar, and also to ascertain beyond question the cause of such hydraulicity. All from the time of Vitruvius had blundered in the dark, working out the old dogmatic form of mixtures without the least knowledge of the action from which they derived their hydraulic properties. The resemblance to the Portland stone, and hence the origin of the name of Portland cement, is clearly also due to Smeaton.

The chemical analyses of the two materials, "Tarras" and "Puzzolana." No. 1, Trass, by Berthier, and No. 2, Puzzolana, from Pit St. Paul, near Rome, by the same authority.

	1.	2.
Silica	57·0	44·5
Alumina	16·0	15·0
Oxide of iron	5·0	12·0
Lime	2·6	8·8
Magnesia	1·0	4·7
Potash	7·0	1·4
Soda	1·0	4·1
Water	9·0	9·2

A great similarity in the value of these two materials is apparent, and the preference given by Smeaton to the puzzolana may be due, in a great measure, to its physical condition or superiority to the trass.

But this painstaking and persevering engineer had not yet completed his investigations, for although he was convinced that, for the purposes of mortars, the combination of the volcanic product with the Aberthaw lime was a success, he had still to test their capacity for being used in a semi-liquid state. We will again follow his own description of this experiment:

"My reader will perceive from the nature of the bond that I proposed in my work that one-half of each piece of stone, being lodged in a dovetailed recess, wherein it was locked fast on three sides, there was no way to get them into their places but by letting them down perpendicularly: in consequence of this, mortar beat up and prepared in a manner similar to what has been specified, could only be applied to the *ground* joint, or *under* bed of each stone: the upright or side joints (supposing the work to be close put together) could not be lined with any quantity capable of filling the whole joint; for if that was attempted, the stone, in being *lowered* into its place, would carry down the mortar upon the sides along with it; leaving so little as not to make these joints solid and full."

In carrying out the experiment to prove the value of a thin grout he put a portion of the prepared mortar into a couple of quart pots (one being filled with beaten mortar), and added water enough to render them so fluid as to enable the mixture to be poured into a mould. They were allowed to stand for a month, and, on the pots being broken, the mixtures were found to be of a *stony hardness*. It was shown that the grout made of the beaten mortar was the best. The experiment was successful, and he says:

"I had therefore no doubt of being able to unite the whole of the materials of my building into one *solid mass of stone*."

At this stage of the experiments, Lord Macclesfield, President of the Royal Society, in a letter to Smeaton, dated London, 14th April, 1757, described what he calls "ash-mortar," and which the bricklayers using it regarded as more valuable for wet and dry work than tarras mortar. It was prepared as follows:

"Take of lime that is very *fresh* two bushels, and of wood ashes three bushels. Lay the ashes in a round trench, and the lime in the middle of the trench; then slake the lime and mix it well with the ashes. Let it lie there till it is cold, and then beat it well together, and so beat it for three or four times before it is used."

On going to Watchet, in Somersetshire, he found that the lias stone from which lime was made there was the same kind of stone as that from Aberthaw. During his travels in this neighbourhood he was surprised to find only one small limekiln at work, and on inquiry found that this lime was unsuitable for agricultural purposes, and a farmer told him that they brought the lime which the land required from St. Vincent, near Bristol, a distance of forty miles. If this lime of the locality (however finely ground) was laid on the land, the first shower of rain would turn it into stone. His observations led him further to investigate the various kinds of limestones in different parts of England, among others:

1. Clunch lime, from Lewes, Sussex.
2. Grey chalk lime, from Dorking, Surrey.
3. Lyas lime, from Long Bennington, Lincolnshire.
4. Aberthaw lime, Glamorganshire.
5. Chalk lime, Guildford, Surrey.
6. Barrow lime, Leicestershire.
7. Sutton lime, Lancashire.
8. Watchet lime, Somersetshire.
9. Berryton lime, near Petersfield, Hants.

In experimenting on the stone from which these limes were obtained he found the muddy deposits, after their having been treated with acid, to be as follows:

1.	2.	3.	4.	5.	6.	7.	8.	9.
$\frac{3}{10}$	$\frac{1}{17}$	$\frac{3}{22}$	$\frac{3}{23}$	$\frac{2}{15}$	$\frac{3}{14}$	$\frac{3}{16}$	$\frac{3}{25}$	$\frac{1}{12}$

It will be profitable for us here to consider the examination made by Smeaton of the effects produced by a combination of plaster of Paris with the mortars. The author is the more anxious to call attention to this point, and in doing so bring to his assistance so eminent and reliable an authority to endeavour to check the tendency to adopt such combinations as novel and useful inventions. Under fresh names many mixtures are sometimes forced into the market, and their adoption secured before it is discovered with what danger and risk their use is attended. Smeaton says:

"Particularly several mixtures of lime with plaster; and although I am aware that the plasterers use some composition of this kind with advantage in their works, yet, when subjected to the test for waterworks, that I had established for the conduct of my own experiments, I found nothing useful in all the mixtures of lime with plaster, nor even of tarras and puzzolana with plaster; as the result was that they rendered the plaster less speedy in setting, and the plaster rendered the compound, that would have resulted from the other ingredients, less firm and more crumbly."

Sir C. W. Pasley is the next prominent experimenter with whose proceedings we shall continue this historical glimpse.

The high position this gentleman occupied as a distinguished officer of Royal Engineers, and his association with all the important scientific bodies of his time, enabled him to command resources of information and facilities for experiment which no other individual less favoured could have secured.

On the threshold of his investigations he ascribes to Smeaton (as we have already shown) the credit of being the first to point out what characteristics in a limestone indicated its claim to hydraulicity, and his ambition was to produce an artificial water-cement from the lines laid down by his eminent predecessor.

His experiments began in the year 1826, by command of the Duke of Wellington, then Master-General of the Ordnance, for the purpose of directing attention to the subject of mortars as a branch of study for the corps of the officers of the Royal Engineers. The first series of these experiments were not satisfactory, owing to his having selected a brick loam to mix with the chalk, and it was not until 1828 that, by a happy accident,* he used the Medway blue clay. In 1830 he considered that the cement he produced was superior in quality to the Roman cement, and indeed its successful application for constructive purposes confirmed his verdict. Before and during his initiatory experiments others had produced equally good cements from analogous materials, and Mr. Frost had actually started a cement factory at Greenhithe, in Kent. We can easily understand the amount of secrecy which would be displayed at the Chatham Dockyard (where Pasley's experiments were conducted), and also the unlikelihood of much information being given from the factory at Greenhithe. Under such circumstances it was not likely that the labours of either could obtain much publicity, more especially when we consider the absence at that time of public interest in the question.

In these experiments he had the advantage of the labours of his predecessors in the line of examination he had entered upon, and, indeed, before they were begun Aspdin had obtained his patent for the manufacture of Portland cement. In the specification of the first patent he is described as Joseph Aspdin, of Leeds, in the county of York, bricklayer.

The specification, dated 15th December, 1824, is as follows:

"My method of making a cement or artificial stone for stuccoing buildings, waterworks, cisterns, or any other pur-

---

* In 1828, my friend and brother officer, Major Reid, having come to reside at Chatham for a few months, chiefly from the interest he took in the architectural course then in progress, requested me to show him not only the mode I had adopted of testing the natural cement and limestones, which had proved perfectly satisfactory, but also in what manner I had attempted to form an artificial cement. I was at first reluctant to repeat any of these experiments, which I told him could only lead to certain failure, but on his expressing a strong wish to see them notwithstanding, I complied with his request; and as my stock of Darland brick-loam was at this time expended, and that place was nearly two miles distant, a different clay was used in lieu of it, which, to our mutual surprise and satisfaction, formed an excellent artificial water-cement.

pose to which it may be applicable (and which I call Portland Cement), is as follows:—I take a specific quantity of limestone, such as that generally used for making or repairing roads, and I take it from the roads after it is reduced to a puddle or powder; but if I cannot procure a sufficient quantity of the above from the roads I obtain the limestone itself, and I cause the puddle or powder, or the limestone, as the case may be, to be calcined. I then take a specific quantity of argillaceous earth or clay, and mix them with water to a state approaching impalpability, either by manual labour or machinery. After this proceeding I put the above mixture into a slip pan for evaporation, either by the heat of the sun, or by submitting it to the action of fire or steam conveyed in flues or pipes under or near the pan, till the water is entirely evaporated. Then I break the said mixture into suitable lumps, and calcine them in a furnace similar to a limekiln till the carbonic acid is entirely expelled. The mixture so calcined is to be ground, beat, or rolled to a fine powder, and is then in a fit state for making cement or artificial stone. This powder is to be mixed with a sufficient quantity of water to bring it into the consistency of mortar, and thus applied to the purposes wanted."

Notwithstanding the existence of this patent, Pasley never alludes to it or Aspdin during the whole period of his experiments. The original process of Aspdin was what we now designate as the "double-kilned" treatment, and with the hard limestones, or their muddy results from the macadamized roads, it was even then considered that a fine condition for admixture with the clays was necessary. The word "impalpability" is made use of in the specification. This process is, however, abandoned, as modern machinery effects almost as perfect a reduction of even the hardest materials as is necessary for their perfect amalgamation with the silica and alumina ingredients.

In 1810, Edgar Dobbs, of Southwark, obtained a patent, dated August 2, which he describes thus:

"The principle of my invention consists

"1st. In making, with the aid of water, a suitable mixture of lime, or its carbonates, with either or several of the following substances, namely clay, loam, mud, slate, road-dust, earth, ochre, cheap metallic oxyd, ores, pyrites, blend, sand, stones, ashes, and all earthy materials (with the exception of lime and its carbonic-acid salts as above) which can be reduced to powder, and which do not vitrify at the heat to which they are subsequently exposed.

"2nd. In expelling the superfluous water.

"3rd. In burning the mixture when hard; and

"4th. In reducing it to powder.

"This powder forms a composition suitable for hydraulic mortar, mortar, stucco, and plaster.

"By carbonate of lime I mean chalk, common limestone, marble, oysters and other shells, earth and earthy materials which slake or fall to pieces after burning, either from exposure to the air or from being brought into contact with water.

"By ashes I mean the natural remains of coals or of vegetables.

"The process is as follows:

"I bring the caustic lime, or carbonate of lime, to a state of fine powder, the former by slaking, the latter by grinding in the same way as chalk, white-lead or flint, or any other equally suitable way, till the particles are so fine that they remain suspended in water, and can be drained off from the coarser particles, which have subsided (and which are ground again); the product is then about as fine as precipitated chalk. I then reduce the other substances, which are to be mixed with the caustic lime, or carbonate of lime, to an equally fine powder. The hard substances are powdered with or without water, and the fine particles removed as above by water; if, however, the substances chosen are too soft to require pulverizing, they are simply comminuted by water, with which they are agitated till it acquires the consistence of a thin paste. I then take a certain quantity of the washed caustic, or carbonate of lime, as the required mixture necessitates, and also of the other washed substance or substances, and mix them well in any suitable vessel. Or the substances are ground or mixed together in their natural state till they become homogeneous and plastic. The mixed substances are then allowed to rest, the clear water is drained off, and the remnant dried either by artificial warmth or by exposure to the air, till the mass is stiff enough to be cut or moulded into pieces before burning.

"The burning can be accomplished in a common limekiln or oven, or even without such an apparatus: but it should be remarked that the intensity and duration of the burning must suffice to consume the combustible parts of the fuel, if contained in the mixture, and to expel the carbonic acid from the lime without vitrifying any of the substances. The burnt fragments are then reduced to powder, either by horizontal millstones, or by any other arrangement for pulverizing dry substances. The composition is then fit for use."

He then goes on to describe the various uses, in combination with sands, &c., for which this lime or cement is applicable.

It is evident that this patentee had a good knowledge of the subject, but he was also afraid of reaching the *vitrifying point*.

We should here, in justice to Vicat, the eminent French engineer, and those who were doubtless inspired by his experiments, refer to a patent granted to Maurice St. Leger, of Camberwell, in May 1818, the specification of which is as follows:

"I take chalk, stone, or any other substance from which lime can be obtained, which I pulverize, to which I add common clay or any other substance containing alumine and silex, which I increase or diminish according to the required strength of the lime. I mix them together, and add water to them until they become a paste of the consistence of common mortar. I then make the said paste or substance into lumps. These lumps, after being thoroughly dried by natural or artificial heat, I put into a kiln, and expose to the action of fire in the usual way of making lime. The degree of heat must depend on the size and quality of the lumps, but I find the lumps have been sufficiently exposed to the fire when they can be broken by the hands. Instead of chalk, stone or such other substances as are above mentioned, ordinary lime, slaked or pulverized, may be substituted; but in that case the compound does not require to be so much exposed to the action of the fire. The quantity of clay, or other substance containing alumine and silex, to be added to the chalk or such other substance as aforesaid, or to the lime slaked and pulverized, must depend as well upon the quality of the chalk, stone, or other substance, or lime, as upon the quality of the clay or other substance containing alumine and silex. But I find in general that from one to two and twenty measures or given quantities of clay, or other substance containing alumine and silex, to every one hundred measures or given quantities of chalk, stone, or other substance, or of lime as above mentioned, is the proper quantity or proportion."

All experiments up to this period apparently aimed only at the obtainment of a resulting product from the kiln which should in its leading characteristics approach the quality of lime freshly burnt. Instead of driving the mixture to the point of vitrifaction, all the operations were so conducted as to studiously avoid such a result; and when by accident this occurred, the pieces so hardened were carefully rejected as worthless.

There were many other patents obtained at and about this time, but they possess no interest and contain but little information to warrant any particular reference to them. While all these patentees and other experimenters in England, France, and Germany were simply seeking, and indeed were apparently satisfied with an artificial hydraulic lime, Aspdin went beyond, and gave the grand finish to the whole by his discovery of the increased temperature of the kiln, and consequent high specific gravity of the cement, now no longer regarded as a simple hydraulic lime.

We are not to suppose that all these various efforts, even including Frost's and Aspdin's, were attended with immediately beneficial results to those who had given so much time to the question, besides the great cost of their experiments. Frost, who was the first to establish works for making artificial cement in 1825, at Swanscombe, in Kent, encountered many difficulties, and for some time after he disposed of these works to Messrs. White, the artificial cement failed to obtain a satisfactory reputation. Pasley considered the proportions then used at these works were inaccurate; at all events until 1833 the artificial cement was selling at one shilling a bushel, and the natural or Roman cement at eighteenpence. Some years later a son of Aspdin commenced to make Portland cement at Northfleet, in Kent, and the quality became more reliable, and since then a continuous demand at home and abroad has led to an improvement in the quality. The profits attending the manufacture of Roman cement were very considerable, and those interested in that trade were naturally opposed to the success of any new cement, however excellent its quality.

Pasley continued his experiments, and tried many kinds of mixtures, but found the most favourable for securing a good water cement to be the alluvial clay of the Medway. This clay, however, he discovered to be unsuitable if allowed to remain exposed to the air for any length of time, a fact now universally acknowledged by Portland cement makers. He also proved that the proportions when mixed and prepared in small balls one inch in diameter for calcination, if allowed to remain for twenty-four hours in the air were deteriorated, and if continued to be so exposed for forty-eight hours would be worthless, whereas balls $2\frac{1}{2}$ inches in diameter were not so prejudicially affected. These are results which experienced cement makers are familiar with in the most modern preparations. We need not follow the experiments with pit clays, slate, fullers' earth, and other ingredients, for they developed no new feature, and resulted in no practical advantage. His experiments with hard limestones were not satisfactory, but, on the contrary, he regarded their treatment for conversion into water cements as a waste of labour, and says, "Good water limes are too common in this country to render it worth while to make artificial ones, and too valuable to be used as an ingredient for artificial water-cement."

The succeeding experiments with chalk and alluvial clay burned in an ordinary limekiln with Newcastle coal produced a weak artificial cement, but one able to hold its own in competition with the then only known water cement (the Roman cement made from the septaria of the London clay). Pasley did not imagine that the preparation required pushing by a higher temperature to the point approaching vitrifaction, or he would have obtained results of a more satisfactory character, for instead of appreciating those pieces which were slightly clinkered by the action of the lime of the kiln, he rejected them as unsuitable and devoid of any cementitious value.

The danger of over-burning the Roman cement was doubtless present to his mind, for any of the over-burnt septaria was useless. The required heat was just enough to reduce the weight of the stone to a minimum, and if the temperature of the kiln exceeded what produced that result the product was unmarketable. Roman cement was best when of the lowest specific gravity.

His mode of testing appears to us as being clumsy and deceptive. It was simply cementing bricks together and exerting a tensile strain until rupture occurred.

It is evident in all Pasley's experiments that he was desirous of obtaining reliable results of an unvarying character, and his disappointments were many, owing doubtless to the want of chemical knowledge or its careful application. The task he undertook was to build up from English materials a water cement so as to equal that of Smeaton's Eddystone in quality. He all but succeeded in doing so, and probably if the duties of his position had permitted him to devote more personal attention to the experiments, the result would have been more satisfactory; but they were spread over too long a period, and were somewhat spasmodic in character. Smeaton, on the other hand, having an immediate and important object in view, ceased not in his endeavours until he had removed from his mind all doubts on the subject of water cements and their qualities. It is true he was assisted in his researches by the fortuitous circumstance of the Italian puzzolana, but it was only his energy and perseverance that obtained from this lucky chance the most advantageous results. The Aberthaw lime in itself could have accomplished all he desired, for he had unlocked the mystery of hydraulicity, and felt confident in the knowledge of its cause. While the military engineer only piled up innumerable disconnected and comparatively useless experiments (from a practical point of view), the civil engineer concentrated his energies and carried his results to a successful end: the one under Government auspices, and the other directed by private enterprise for the accomplishment of a great work of human benevolence and universal importance. The Eddystone Lighthouse still stands to testify to the great ability of its constructor, and should be regarded by both makers and consumers of Portland cement as a beacon to light them in the way to its true and honest use.

The Portland cement manufacture languished for many years, and indeed in this country at least it ran the risk of being utterly extinguished owing in a great measure to the hold Roman cement had obtained of the constructive profession. This long-established rival, however, gradually succumbed to the merits of the new product, and the impetus given to its use by the London Drainage Works in 1859 settled the question of its superiority, and it may safely be asserted that the quality now obtainable is 100 per cent. better than that made before that time.

Some years after Smeaton's experiments, but before the publication of his great work, 'The Narrative of the Eddystone Lighthouse,' in 1791, Dr. Higgins published in 1780 his 'Experiments and Observations on Calcareous Cements.' He was previously prepared for the task he had undertaken by the information acquired from the writings of the celebrated Dr. Black (the discoverer of carbonic acid gas), and in the following words acknowledges that indebtedness:

"I had already learned from the chaste and philosophic productions of Dr. Black that calcareous stones which burn to lime contain a considerable quantity of the elastic fluid called fixable air or acidulous gas, which in combination with the earthy matter forms a great part of the mass and weight of these stones; and that the difference between limestone or chalk and lime consists chiefly in the retention or expulsion of that matter."

Dr. Higgins first entered on his experiments to ascertain

at what temperature the carbonic acid gas could be expelled, and the time such expulsion occupied. They were laboratory tests, and are chiefly remarkable for the care with which they were conducted, and the sensible conclusions arrived at by the experimenter. He obtained "letters patent" for his invention of a water cement or stucco made from prescribed admixtures of sand, rubble, lime, and bone ashes, and it would appear at that time this cement was received with much favour, and used by architects and builders of eminence. The process of preparation, however, must have been tedious and expensive, and could not withstand the competition and rivalry of Parker's cement, for which the patent was granted in June 1796. It is just possible that if Dr. Higgins had been an engineer and required cement for his own work, he would have found out what had been done by Smeaton, and directed his attention to the addition of clay to the chalks or their limes instead of the fabrication of a compound mortar of doubtful value. Hitherto it had been well known from the writings of Vitruvius that the Romans used fat limes in combination with various volcanic products, and the accident of research after good mortars for the Eddystone structure disclosed the existence of hydraulic limes. At this time the limited use of mortars conduced to the subject remaining in a state of comparative darkness, but the increasing wealth resulting from industrial pursuits gave an impetus which necessitated the construction of great engineering works.

Ancient and extinct civilizations had successfully used cements and mortars, as the existing remains of their engineering and architectural works fully testify. These remains of a varied and interesting character even now show us how much of their stability and permanence is due to the quality of the cementing agent by which the building materials were put together. Much admiration has been expressed on the character and quality of ancient structures, and puzzling guesses as to the mode and manner of the preparation of the mortars. Lime mortar as a binding material was used by the Phœnicians in the island of Cyprus, as shown by the temple ruins of Lanarca. The Egyptians in the Nile Valley are supposed to have made use of the river mud and burnt gypsum as a mortar in building the Pyramids. The Assyrians had recourse to asphalte as their cementing agent in the construction of Babylon and Nineveh. The asphalte of the Dead Sea was well known to the ancients. It was also obtained in pits and springs near the rivers Euphrates and Tigris.

Even with such ancient testimony it is difficult to say when and where the first mortars were used. The term or name "mortar" is of Roman derivation, from the fact of these thoroughly practical people using a "mortarium" to secure the perfect homogeneity of the binding material. The name therefore should now be regarded in a generic sense, and alike applicable to all preparations used in binding together in structural form stones, bricks, or other building materials.

While referring to these examples of ancient cementing agents, we are tempted to contrast the position of the modern Portland cement, which penetrates to every accessible quarter of the globe where engineering science demands its aid. From Russia to India in the old world to the North and South Americas of the new, "English Portland cement" is an article of universal currency, and almost becoming one of necessity. The port of London attracting the commerce of the world to its docks and river, naturally assisted in the establishment of a belief that Portland cement was indigenous to its neighbourhood, and could only be produced by a combination of the chalks and clays found on the banks and shores of the river Thames and its affluent the Medway.

In France and Germany, however, large works have been for a long time in existence, and it may be said that London, Boulogne, and Stettin respectively represent the chief centres of this industry in these three countries.

From the preceding remarks it must be obvious that a difficulty occurs in attributing to any one of the numerous experimenters the credit of inventing Portland cement. We find differences, however, in the value of the assistance rendered at the various stages of progress, and Aspdin, although the least distinguished of all of them, persevered in bringing to our knowledge the importance and, indeed, necessity of a high temperature in the kiln. His sole aim was to render his supposed invention profitable, and he ultimately paved the way for a more extended system of manufacture by others in more favoured localities than that wherein he fixed his own isolated works. We have in this circumstance and its results ample evidence of the necessity of adapting any novelty to the wants of the public if it is wished to become ultimately a profitable commercial success.

Having said this much of the introductory or historical features of our subject, we will now shortly state what Portland cement really is, for in the somewhat, and perhaps necessarily, confused accounts of its beginning this has been overlooked.

Portland cement is a chemical product obtained by a preliminary mechanical combination of carbonate of lime with silica and alumina, which, after passing through the succeeding chemical stage of manufacture may be described as a double silicate of lime and alumina. The name given to it by Aspdin has tended to some confusion of ideas about its source and origin, many supposing it was the product of converted oolitic limestone from the island of Portland, in Dorsetshire. A close resemblance, when of an unexceptional quality, in colour and texture to the celebrated Portland building stone favours this belief. This cement is simply an improved hydraulic lime.

The term hydraulic is borrowed from the French engineers, who used it to distinguish the limes capable of setting under water from the common limes devoid of that property. It is synonymous with the term "water," which English engineers formerly applied to all limes used in the construction of sub-aqueous masonry. This distinctive term "hydraulic" is now generally given by European engineers to all cements or limes possessed of that property, whether artificial or natural.

Vicat was the first, probably taking the idea from Smeaton, to experiment on the possibility of rendering the rich (common) limes hydraulic by mixing with them (both in their carbonate and hydrate state) various proportions of different kinds of clays. When these experiments (probably the earliest reliable ones that were guided by chemical reasoning) were first initiated, considerable doubt existed as to the true source from which the faculty of hydraulicity was derived; some authorities maintaining that this valuable property was due to the oxide of manganese, and others attributing this effect to the existence of the oxide of iron. After a great deal of controversy, however, it is now finally conceded that silica and alumina, but especially silica, imparts to rich limes the desired quality of hydraulicity; the presence of the oxides of iron or manganese within certain limits being purely negative in character, so that in its simple and clearly defined position of an improved artificial cement the merits of this material should be easily understood.

There is still a good deal of mystery imported into its manufacture, damaging to those who persist in its continuance. The author, while paying a recent visit to the parent or original manufactory at Wakefield, experienced the still existing tendency to enshroud the process of manufacture with an air of ignorant exclusiveness.

The most valuable property perhaps which a pure Portland cement possesses is that of being invariably uniform in quality. Such a result is, however, only attainable under accurate manipulation in every stage of its manufacture. It is this particular property, more than any other, which has tended most to its advancement, and which causes it to be especially acceptable to engineers for large and important works of all kinds, both on land and in water. This gratifying position could not have been reached if engineers had neglected their duty of rigidly testing the cement before it was used. Manufacturers now understand the requirements of engineers, and act wisely in meeting them in such a way as to increase the good reputation of their manufactures.

In the case of natural limes and cements considerable difference of value arises in the products from the kilns, owing to the variable character of the mineral deposits of the quarries. Such objection applies more especially to the marls and clays of the lias formation, and, indeed—except in some rare instances—the production of a first-class Portland cement, by one direct and simple operation of calcination, is unattainable. It is seldom that any natural cements are capable of satisfactorily undergoing the ordinary tests applied to a high-class artificial Portland cement. In the valuable property possessed by Portland cement of retaining its cementitious value for a lengthened period of time, they are also deficient; and unless natural cements, of any class, are exceptionally good, they may be regarded as of less constructive value than the hydraulic limes of the blue lias and other analogous formations.

The gradual increase in the use of Portland cement for

large engineering works has led, in a corresponding degree, to a diminution of that of hydraulic limes and natural cements. Where large quantities of hydraulic limes are still used in important works much care is bestowed in securing them freshly burnt and finely ground. Under such precautions they may be profitably mixed with the various sands and gravels in the preparation of mortar and concrete.

Any description of the various characteristics of Portland cement would be incomplete if, among its many advantages, we were to omit its most salient property of comparatively quick setting. This is a property, however, possessed by nearly all the natural cements, although their setting differs considerably. In the case of natural cements it is rapid, and the maximum induration quickly attained; whereas in the case of Portland cement, of the highest class, the setting process is protracted; and, indeed, for its most important uses, only such cement is admissible.

Portland cement not only commands an important position as an English manufacturing product, but as an article of export it occupies a prominent place. It is not to be supposed, however, that this is due to any want of producing capacity on the part of either France or Germany, for near Boulogne in the one country, and at Stettin in the other, there exist, perhaps, the largest Portland cement works in the world.

The home consumption is very large indeed, yet it is much exceeded by the amount exported to nearly all parts of the world. Many of the most important engineering works in France and Germany have been executed with English cements, although, as before stated, these countries can produce for themselves. Our extensive export trade is in a great measure due to our commanding shipping facilities, which is a matter of great importance in dealing with a heavy article like Portland cement of comparatively low pecuniary value.

We may briefly give the chronological order of the experiments so as to show how much ingenuity had been expended before the realization of the Portland cement desideratum was reached.

First. Smeaton in 1756, to obtain a special cement for his own works. Second. Higgins, in 1780, for the illustration and development of his particular views and theories of a cement more especially suited for stuccoing or house-fronting purposes. Third. Parker, in 1796, with the most fortunate and useful discovery of converting the nodules (septaria) found in the London clays. Although we may now be able to discard the cement derived from this source, having at our command one of much greater value, it served most useful purposes. The Thames Tunnel could not have been made but for the advantages it secured, and many of the early railway tunnels were built with it as a cementing agent. Fourth. Dobbs, in 1810, shows by his specification that he had a good mechanical knowledge of the subject, but his chemistry was insufficient to guide him. Fifth. Foreign experimenters, as Vicat, John, Treussart, and St. Leger, all of whom aimed at the obtainment of an artificial hydraulic lime during and about 1818, assisted in an able and satisfactory manner with their chemical knowledge in the elucidation of the subject. Sixth. Aspdin is entitled to the next position, from the date of his patent, in 1824. Seventh. Pasley, for many years after 1826, when he first entered seriously on the subject in the extensive experiments which he so ably conducted. Eighth. Frost, in 1826, from his being the first to erect a manufactory, near London, for the purpose of making Portland cement for constructive purposes.

Before closing this chapter, we in the most hearty manner

acknowledge the great services rendered to the scientific branch of the subject by Captain Smith, by his valuable translation of Vicat's work on mortars and cements, published in 1837 : a careful study of which will greatly assist the intelligent investigator in his future and more advanced studies and experiments.

CHAPTER II.

GEOLOGICAL AND MINERALOGICAL OBSERVATIONS.

Natural Sources from which the Raw Material may be obtained.

THE most important ingredients required for the fabrication of a good Portland cement, when truly combined, are *carbonate of lime, silica, and alumina.* It is seldom, however, that these leading and indispensable minerals are found in such a state of chemical purity as to be accepted without careful analysis. The impurities with which they are generally in combination, such as oxides of iron, manganese, magnesia, &c., are not up to a certain defined limit dangerous, but act simply in a negative capacity, and in such condition are harmless in character.

Aspdin in the original or *parent* Portland cement manufactory adopted those materials which the locality commanded, and with whose properties he was acquainted. They were, however, unsuited, as well as the site of his works, for an extensive system of manufacture, and other localities were sought for where the necessary conditions of success could be more readily commanded and secured.

London was chosen as the proper centre for the establishment of the new manufactory, commanding as it does all the required conditions for success; a great commercial centre in and around which the new product could be absorbed in the numerous works of construction, as well as having within a convenient distance inexhaustible deposits of the most suitable raw ingredients. The chalk formations on the banks of the rivers Thames and Medway, and the deposits of mud in their estuaries and creeks, furnished at an inconsiderable cost the most suitable supplies in the best and most acceptable form.

Fuel was readily obtained in the shape of coke from the gasworks of the metropolis, and as a crowning advantage to the happy combination we have described, the port of London, with its numerous shipping facilities, afforded the means of exporting the cement to distant countries on the most advantageous terms.

These favourable and singularly fortuitous circumstances resulted, as might have been expected, in the establishment of extensive cement manufactories in the neighbourhood of London, and it is now regarded as the centre and largest field of this industry.

Such a favoured combination naturally placed the London cement makers in a position of commanding influence, so as to secure for them a practical monopoly of the cement trade, not only of the home trade, but the foreign trade as well. Hence the hitherto prevailing notion that Portland cement was only made in London, and the materials for its manufacture were elsewhere unattainable.

Before entering on the discussion of the various cement-making materials, we are desirous of calling attention to the desirability of inculcating in the minds of the operative the advantage, if not the necessity, of acquiring something more than a simple external or surface knowledge of the materials passing through his hands, and in the conversion of which he plays so important a part. In other industries besides that of cement making the ignorance of the manipulator debars him from the enjoyment which a knowledge of geology and mineralogy would afford. Why should the sciences allied to the industries of the coal miner or iron worker remain a sealed book to them, and beyond their intelligent comprehension? Is it creditable in this age of intelligence that the operative cement maker while dealing with the simplest of minerals should only exert mechanically the aid required of him, and continue debarred by his ignorance from appreciating the nature of these materials and the original sources from which they were derived.

It may not be possible, nor is it necessary, that the workman should have his mental training so acute as to be capable of determining like a De la Beche from the physical aspect of a country its geological period, or its mineral value with the facility or unerring instincts of a Dana, or its chemical value like a Liebig. Such prescience is only within the grasp of specially trained and highly gifted intellects.

"Hard work" is made harder when pursued without interest. It is the wearisome uncongenial task that ultimately breaks down the most elastic mind.

There is amongst several classes of operatives certain acquired knowledge resulting from intelligent observation in their pursuits, which assume practical application according to the sagacity of the observer. A familiar example of this kind of instinct is shown by the Scotch coal miner when deciding on the character and direction of the dislocation of the seam of coal, when in "troubled fields" they are intersected by trap or volcanic dykes or faults. The test indication of the "coom," the remains of soot caused by the igneous action of the molten trap at the period of its intrusion, showing by its appearance whether the dislocation has led to the "upthrow" or "downthrow" of the seam of coal.

The French quarryman, to give another simple and familiar example, in quarrying the famous Caen stone, estimates its quality by a kind of musical scale expressed by the application of his hammer. The sounds thus developed are *pif, puf, pouf.* The first indicating the hard, and the last the soft and friable.

Such rules are somewhat dogmatic in character, receiving their value only by the amount of intelligence exercised in their application; at all events they fall far short of the unerring rules prescribed by science.

Analyses of Caen Stone.

	1.	2.
Water	1·91	1·28
Silica	13·71	13·53
Protoxide of iron	0·73	0·72
Carbonate of lime	82·58	82·38
Magnesia	0·48	0·52

Not only is the attainment of useful rudimentary knowledge possible, but, as we will show, the higher rounds of the scientific ladder have been reached by the humblest individuals. In the field of geology two remarkable examples are to be found in the persons of Hugh Miller and Mary Anning; the former, of world-wide reputation as an original discoverer and thinker, and the other probably as well known to the leading geologists of her time from her diligent labours in the exhumation of antediluvian animals; the one devoting his great natural abilities to the development of a knowledge of the geology of his native country, and the other assiduously, in a not less useful manner, prosecuting her researches after the long-buried remains abounding in the vicinity of her birthplace; both alike imbued with strong religious feeling, which must have been intensified from their capacity to translate or read the great book of nature, and the wondrous works therein recorded of its Author.

A short account of the circumstances which originally led

to the connection of these two celebrities with the science of geology is peculiarly interesting.

Mary Anning, the daughter of a carpenter, was left an orphan at eleven years of age, and at that early period commenced seeking for strange forms along the sea-beach in the neighbourhood of Lyme Regis for a livelihood. Her first find she sold to a lady for half a crown, and this encouragement induced her to further and more diligent research. A few months afterwards she saw in the lias strata the projecting bones of an animal, in the digging out of which she employed the required manual labour. The skeleton thus obtained she sold for 23l., and is the ichthyosaurus now in the geological gallery of the British Museum. It is not to be supposed that this discovery led to the immediate classification of these bones so singularly and by such humble agency brought to light. It was not until ten years afterwards that the illustrious Cuvier finally allotted them their palæontological position. Through these discoveries Mary Anning became famous, and corresponded with Home, Buckland, Conybeare, De la Beche, and Cuvier, and they, as well as numerous students of geology, were indebted to her for many of their illustrative specimens.

Here, then, is a good example of what may be accomplished by the humblest artisan or hand-worker, and shows how even a comparatively ignorant and obscure orphan girl became the correspondent and associate of illustrious and eminent men. Indeed, her fame became widespread, and fully justified her humble claim to notoriety, and she may well be pardoned the egotism expressed in her reply to the King of Saxony, when she said, "I am well known through the whole of Europe."

We will let Hugh Miller speak for himself out of his 'Old Red Sandstone.'

"I set out a little before sunrise to make my first acquaintance with a life of labour and restraint, and I have rarely had a heavier heart than on that morning."

In the laborious occupation on which he had entered he met with the unavoidable hardships with which it was associated. But after a short interval he found ample recompense for all the discomforts he had endured. The first quarry at which he laboured for a short time was abandoned, and in another one in the Bay of Cromarty, situated in a highly interesting geological formation, he first felt that inspiration which eventually led him to a high position in the scientific and literary world. In describing the locality in which his labour duties had so happily placed him, he says:

"In short, the young geologist, had he all Europe before him, could hardly choose for himself a better field. I had, however, no one to tell me so at the time, for geology had not travelled so far north; and so without guide or vocabulary I had to grope my way as I best might, and find out all its wonders for myself. But so slow was the process, and so much was I a seeker in the dark, that the facts contained in these few sentences were the patient gathering of years.

"In the course of the first day's employment I picked up a nodular mass of blue limestone, and laid it open by the stroke of a hammer; wonderful to relate, it contained inside a beautifully finished piece of sculpture, one of the volutes apparently of an Ionic capital; and not the far-famed walnut of the fairy tale, had I broken its shell and found the little dog lying within, could have surprised me more. Was there another such curiosity in the whole world? I broke open a few other nodules of similar appearance—for they lay thickly on the shore—and found that there might. In one of these there was what seemed to be the scales of fishes and the impressions of a few minute bivalves, prettily striated; in the centre of the other there was actually a piece of decayed wood. Of all nature's riddles, these seemed to me to be at once the most interesting and the most difficult to expound. I treasured them carefully up, and was told by one of the workmen to whom I showed them, that there was a part of the shore two miles farther to the west where curiously shaped stones, somewhat like the heads of boarding pikes, were occasionally picked up; and that in his father's days the country people called them thunderbolts, and deemed them of sovereign efficacy in curing bewitched cattle. Our employer, on quitting the quarry for the building on which we were to be engaged, gave all the workmen a half holiday. I employed it in visiting the place where the thunderbolts had fallen so thickly, and found it a richer scene of wonder than I could have fancied in even my dreams."

It is unnecessary for us here to follow further the interesting history, but would recommend a reference to the work itself ('Old Red Sandstone'), from which may be derived much delightful scientific knowledge, imparted in the most charming and acceptable manner.

If it were desirable we might furnish many more illustrations of the assistance afforded to the highest trained intellects by the most humble agencies. We cannot leave this interesting page of our subject, however, without mentioning the widely-known assistance or help afforded to the indefatigable naturalist, Mr. Frank Buckland, by his humble fishermen friends, who secure for him rare specimens for the illustration of the science of which he is so able and original an expositor. Theirs is, indeed, a labour of love, attracted to the person of the genial recorder of nature's living mysteries in their most interesting and alluring forms.

Before entering on the more dry and scientific part of our subject we would preface the introduction thereof with the observation that exception may be taken to the prominence which we give to the geological branch, and some may even question its relevancy. But we hope, before closing our labours, to show that in cement making its importance is only subordinate to that of chemistry. Geology has done much for agriculture: it developed the use of coprolites; gypsum through its agency was applied to the improvement of the growth of clover. The danger of using, for agricultural purposes, lime made from the dolomite or magnesian limestones, was averted, and much valuable knowledge was acquired from its application in arriving at the value of the soils and the kind of manures best suited to their various characters.

We do not intend to refer, in exact scientific terms, to the various geological deposits from which the raw materials for cement making are to be obtained, but discuss them in such order as they appear to us requisite to the subject, giving them priority as we measure their value and importance from a cement-making point.

Carbonate of Lime.

This mineral is one of the most abundant constituents of rocks, rendering it almost impossible to define, with any degree of accuracy, the limits of its all-pervading presence. It occurs but rarely in a state of purity, and is only represented in that condition by calcspar and arragonite. In chemistry it is known as the metal calcium, discovered by Sir Humphry Davy.

Pure carbonate of lime consists of

Lime	56·3
Carbonic acid	43·7
	100·0

Carbonate and sulphate of lime have their representatives in nearly every stratified formation. The marbles of the *Metamorphic*; the corals, or shells, of the *Silurian*; the coralline and shell marbles of the *Devonian*; the encrinal shells and fresh-water beds of the *Carboniferous*; the dolomites of the *Permian*; the lias of the *Triassic*; the oolites of the *Jurassic*; the shelly bands of the *Wealden*; the chalks and gaults of the *Cretaceous*; the nummulites of the *Tertiary*; and the plastic clays of the *Post Tertiary*.

In the various estimates made of the extent of the limestone formation it is computed that in one or other of its varieties it occupies three-fourths of the earth's surface. England and Ireland are especially well provided, but in Scotland it is less plentiful. Scotland's proximity to Ireland, however, secures for it a plentiful supply of lime for building and agricultural purposes, in exchange for coals, which in that country are comparatively abundant.

The limestone tract in Ireland is of considerable extent, having a range from east to west, or from Dublin to Galway, of 120 miles, with a mean breadth of something like 100 miles. It seldom exceeds 250 feet in altitude above sea level.

Carbonate of lime is insoluble, or only very slightly soluble in pure water, requiring 778 grains of water at a temperature of 60° Fahr. to dissolve one grain of lime, and at boiling point 1278 grains. In a hydrate state it is much more easily dissolved.

According to Fresenius, one part of carbonate of lime can be dissolved in 8834 parts of cold and 10,601 parts of boiling water. When rendered caustic, as proved by Bineau's experiments, it is soluble in the proportion of 780 parts of cold and 1500 parts of hot water to 1 part of caustic lime.

This extreme insolubility of the carbonate of lime does not affect the transparency of the water in which it is dissolved, and hence the necessarily slow and protracted process of its deposition in the various processes with which in the geological system of our globe it has played, and still continues to play, so prominent a part.

Notwithstanding this extreme minuteness of solution in the water, it is found to interfere prejudicially when so combined in the water used for domestic and manufacturing purposes. This difficulty is, however, readily overcome by the application of Dr. Clarke's process as now adopted in recently constructed waterworks when the supply is pumped from the chalk formation.

The white crystalline marbles contain but a slightly appreciable quantity of impurities, as the following analyses by Wittstein indicate:

No. 1, specimen from Carrara.

No. 2, from Schlonders, in the Tyrol, crystalline.

No. 3, from the same locality, compact.

	1.	2.	3.
Specific gravity	2·732	2·700	2·566
Carbonate of lime	99·236	99·010	97·040
,, magnesia	0·284	0·521	2·109
Oxide of iron	0·251	..	0·360
	99·771	99·531	99·509

It may here be observed that mineralogists are not quite agreed as to the accurate application of the term "crystalline." Our use of that term may only, therefore, be regarded as simply distinguishing the physical texture of the rocks, and not in their strict or accurate mineralogical sense. Such a classification as the following will be observed, viz.:

Perfect crystalline, as in calcspar; minute crystalline, as in the fine-grained marbles; sub-crystalline, as in the oolite or the powdery and amorphous state of the chalk.

It is believed that volcanic action has altered, in many formations, the original character of the carbonates of lime long subsequent to the period of its primary deposition.

Numerous examples of this metamorphic change are to be found, amongst others that of the Carrara marbles, which, originally oolite limestone, became by this action the most valuable of known statuary marbles. In Ireland again, at the famous "Giant's Causeway," the intrusion of the molten basalt converted the chalk in the neighbourhood of its occurrence to a granular condition, thus destroying or changing its original or amorphous character. In North Derbyshire we also find the carboniferous limestone converted into a variety of marbles, almost of every shade and colour, by the extrusion of the igneous material known in the locality as "toadstone."

These several metamorphic changes must have been produced when the extruded material was at a very high temperature, and under conditions of enormous pressure, otherwise the carbonic acid gas would have been evolved, and instead of crystalline marble there would have been lime.

From such widespread and, comparatively speaking, inexhaustible supplies of carbonate of lime, the cement maker need find no difficulty in obtaining this, the first and most important ingredient for his manufacture. Besides the interest and value attaching to this material from a cement-making point of view, it has other claims on our attention from the singularly strange and varied processes by which it was formed.

We will offer a few remarks on the generally accepted theories of the origin and growth of the earth's crust.

Since the discovery of spectrum analysis, and its application to the purposes of astronomical observation, observers have concluded that the earth was originally in a gaseous state, at which time calcium, carbon, and oxygen existed in a free uncombined state on account of the then prevailing high temperature. On this temperature lessening carbonic acid was formed, and the calcium combined with oxygen, forming lime. At this period the temperature was still too high for water to assume a liquid form. Experiments of an interesting and reliable character, made by Schulatschenko, demonstrate that carbonic acid cannot combine with lime in the absence of water in its normal or liquid state; it follows that carbonic acid and lime co-existed at this time without being in a state of combination. The lime, however, was already in contact with silica—a much more powerful agent at high temperature than carbonic acid—so that where the gradual lowering of the globe's temperature resulted in solutions of carbonic and other more powerful acids, they combined with silicate of lime and various igneous compounds. These were the more easily decomposed from the then existing enormous pressure caused by the quantity of water in the altering atmosphere. Whether under this pressure carbonic acid had previously existed in a liquid state is difficult of proof, as we are without experiments to explain the action of liquid carbonic acid on lime and its silicates. To have rendered possible the existence of liquid carbonic acid a pressure of 36 of our atmospheres would have been necessary. Silicates of lime are decomposed by carbonic acid; the lime found in the springs of the granite formation is due to this source.

Assuming that these theories are tenable, for at present there is no information or research which disturbs such conclusions, we arrive at the reasonable inference that all the carbonates of lime are the result of decomposed pre-existing silicates. In modern times we have no evidence of any natural process by the agency of which silicates of lime are formed equalling in quantity those decomposed by water containing carbonic acid. There is no example where a deposit of carbonate of lime can be proved to have been deposited by igneous action without subsequent metamorphosis. Even the old crystalline marbles in the gneiss of Rio Janeiro are so clearly stratified that Mr. W. F. Reid, during his recent visit to the Brazils, was enabled to point out to a local lime-burner the continuation of a stratum apparently lost from its partial obliteration by erosion.

If the surface of the globe were covered with water, and a hot solution of bicarbonate of lime injected therein, there would be no precipitate of carbonate of lime owing to that substance being more soluble in cold than hot water.

The above explanation of the primordial action during the earth's infancy brings us to the period before the "dawn of life," and in the absence of all organisms. We shall now examine the processes which are supposed to have continued the wondrous operations during the succeeding geological epochs, and their present and continuous action in changing the aspect and conditions of modern geology.

In the sub-structural rocks, formed before and partly during the early Devonian period, carbonate of lime occurs in comparatively insignificant quantities. Where and whence, then, the sources from which the materials were obtained and used in building up the vast deposits of carbonate of lime? The explanation, or answer, is sought for in the following manner.

In the fractured conditions of the earth's early crust numerous calcareous streams, forced by enormous internal pressure, issued through the crevices. At this period water covered nearly the whole surface of the globe, and thus became charged with the ejected and boiling solutions of bicarbonate of lime, magnesia, &c. These agencies had their origin in a period so remote as to baffle our powers of computation to determine their beginning or end. The ocean, therefore, during all past time, and indeed at the present time also, became the storehouse, supplying the varied vitalities with the soluble materials to construct and maintain their organisms, which by their decay eventually formed the deposit of carbonate of lime. Such agency may be considered a chemical one, but there are others of a mechanical character ceaselessly at work in changing the face of nature, and by their action contributing to the building up of new mineral deposits. Of this class of agency the mechanical force of river action is not the least important in its effect and results. While depositing in its strictly river course vast quantities of insoluble materials, it carries in its seaward course others of a more soluble nature to be first absorbed in the various oceans of which they are the feeders, and at last utilized in the growth of organisms before referred to. The amount of river water flowing into the various seas and oceans is something enormous. We shall shortly allude to a few of the best-known examples. The Rhone, through its 500 miles drains an area of 7000 square miles. The Rhine, with its 800 miles, conveys to the sea the drainage of 14,000 square miles. The Danube, 1800 miles long, accommodates the drainage of 55,000 square miles of country. In North America, the St. Lawrence, traversing a distance of 1200 miles, discharges into the Atlantic Sea the drainage of 300,000 square miles. The great Mississippi, meandering through its tortuous course of 3800 miles, ultimately conveys the drainage of 1,000,000 square miles of country.

It is estimated that 1,800,000,000 tons of water enter into the Mediterranean Sea during the twenty-four hours, being the collective contributions of all great and secondary rivers discharging along its shores. The Black Sea receives the discharge of one-fourth of the river waters of Europe. The flow of the river Thames at Teddington on a rough average is 8332 gallons of water per second, a gallon of which yields on analysis $22 \cdot 5$ grains of solid matter, chiefly bicarbonate of lime, equal to $96 \cdot 417$ lbs. per hour, or 377,058 tons per annum.

The river Danube, after its traverse through 1800 miles

of fertile country, discharges itself into the Black Sea. During floods, and when its waters are most surcharged, they bear to the tideless sea as much sedimentary matter as, if precipitated and compressed, would be equal to 1 cube inch of solid matter to a cubic foot of water. Not above one-fortieth of this enormous quantity is, however, transported when the highest floods have subsided. Upwards of 600,000 cubic yards of diluvial detritus pass into the sea by the several mouths of the river in twenty-four hours, and during its normal state 15,000 cubic yards in the same time. The maintenance of a navigable mouth has been successfully accomplished, through the skilful agency of Sir Charles Hartley, to whom the European Commission entrusted its execution.

The Nile again, which through all historic ages has been the wonder and admiration of both learned and ignorant: the mystery of its source, and the marvellous fertility resulting from its periodical inundations, combined to render it, from the earliest period of recorded history, a subject of much controversy and speculation. The solid matter annually brought down by this river has been estimated, in pursuance of very careful measurements carried on for many years, at 240,000,000 cubic yards, that is, equal to an area of 2 square miles 50 feet deep. It is thought that a portion of this mass is carried eastward by the prevailing shore-current that sweeps the fringe of the Egyptian Delta, and either raises the bed of the Mediterranean or reappears in the shape of blown sand creeping over the gardens of Palestine. But, all deductions made, a steady annual growth of cultivable land is added year by year to the dominions of the Khedive of Egypt.

All these mighty masses, thus transported by natural agency in never-ending and continuous regularity, so far eclipse by their immensity any power at the command of man, or under his control, that even the combined energy of all the steam power on the globe exerted to its utmost point would, in comparison, be of the most pigmy character.

The degradation of the various geological formations, due to the various atmospheric and meteoric agencies, assists to a great extent in preparing for the transport of these vast quantities so various in mineral character and value.

These illustrations form but an insignificant item in the large total of the river power of the globe, but they give us some idea of the force of such agency, and its ceaseless contribution to the ever-changing geological conditions of our planet.

Wells and springs, by the deposition of the salts in their water, assist also in the production of carbonate of lime. A familiar example of this class of agency occurs to us in the case of the Bath old wells. A daily water yield of 181,440 gallons yields 608 tons of solid matter every year.

In Italy a building stone, known as travertine, is composed solely of carbonate of lime deposited from volcanic waters. Some springs in Tuscany deposit as much as 12 inches of carbonate of lime during the year. Similar springs occur in England, as at Matlock, in Derbyshire, and other places, where they are locally known as petrifying wells. Carbonate and sulphate of lime is, indeed, to be found in nearly all wells, where its presence is indicated by the deposited incrustations in cooking vessels, which is simply lime, or its silicate, after being deprived of its carbonic acid. Water in the London chalk district contains 20 grains of bicarbonate of lime in every gallon.

The celebrated "Falls of Niagara" afford a favourable illustration of another kind of agency. The abrading action of the river flowing over the falls has reduced, during fifty years, the platform of the cataract to the extent of 40 yards, or nearly 1 yard per annum; the degradation of the old rock resulting in the formation of a soft shelly limestone in the new or lower river bed. The distance from the lower opening of the narrow gorge to the present face of the cataract is 7 miles, and the time occupied in this process of retrogression was computed, by the late Sir Charles Lyell, at 35,000 years: that is, from the period of the drift, or glacial action, to the present time. The falls are now distant 25 miles from Lake Erie, and when that space has been traversed by this destructive hydraulic agent the lake itself will be emptied, from the fact of its depth being less than the height of the falls, unless during that period the volume of water should become diminished and its force thereby be prevented from maintaining the present height of the fall.

We will now proceed to describe the various formations from which supplies of carbonate of lime may be obtained. The first to which we shall call attention is, perhaps, the best known and most valuable, and with which, in England, we are best acquainted, viz.

CHALK.

Chalk is a fine granular carbonate of lime, existing geologically in three well-defined formations known as

Chalk Marl,
Lower Chalk,
Upper Chalk,

the latter being frequently interspersed with layers of flint nodules.

It occupies a large portion of England's surface, and if the thin soils, with which in some places it is covered, were bared, it would be traceable in varying breadths from Dorsetshire to Flamborough Head, in Yorkshire. With the exception of the wealds of Kent and Sussex it prevails in all the south-eastern counties.

Notwithstanding the extent of the English deposits, and even after adding those of the north of Ireland, they form that the cement maker is to apprehend danger, from the consequences of which he can only protect himself by the exercise of the most careful attention.

The next deposit we purpose considering is the lias, which we regard as second in importance—from the cement-making point of view—to that which we have just referred to.

Lias.

This formation is so named, it is supposed, from a corruption of the word lyas, or layers, and was originally applied to the thin-bedded limestones at the base of the oolitic system. It is generally arranged in the following divisions:

1st. Upper lias clay, or shale.
2nd. Marlstone.
3rd. Lower lias clay, or shale.
4th. Lias rock.

Although, perhaps, not equalling the chalk in geographical extent, it, however, occupies a considerable portion of this country's surface, and extends in almost unbroken continuity from Lyme Regis in the south, to the river Tees in the north, with isolated patches in North and South Wales and elsewhere. In Scotland and Ireland it exists but in comparatively insignificant quantities. The most favourable points for studying and observing this highly interesting geological formation are Whitby, in Yorkshire; Westbury, in the estuary of the river Severn; Watchet, in Somersetshire; Aberthaw, in Glamorganshire; Lyme Regis, in Dorsetshire; and Barrow, in Leicestershire.

Besides its industrial value this formation commands much attention from its being the depository of the gigantic fossil remains of the ichthyosauri, plesiosauri, and other extinct animals. Its prevailing and distinguishing colour is blue, its shade varying in different districts, even reaching that of a pale grey or white, known in some localities as the "white lias," when it assumes a highly indurated character, and when in that state is used for ornamental and other constructive purposes.

This formation, from its muddy appearance, and the almost entire absence of any crystalline characteristics, induces a belief that some different natural process was at work than that which built up the chalk and other carbonate of lime deposits. The disproportion of the limestone to the shale is very marked, and although the former is seldom found with a larger proportion than 80 per cent. of carbonate, and then in comparatively shallow beds, the clays preponderate, and even in the best circumstanced quarries, unless where brickmaking is carried on, large quantities are wheeled to waste, thereby involving an increase in the cost of working.

The perfect animal remains found in the "lias" induces the belief that the growth of its deposit was at a rapid rate, the unconsumed food, and other conditions of the exhumed remains, favouring this supposition. That the deposit was performed in still and placid waters is beyond question, from the accuracy—even to the colour in the shells—of the most delicate remains. The ooze bed of the great liassic sea formed a soft resting place for the delicate pentacrinites, as well as the huge saurians, both of whom, according to their weight, became entombed, to be recovered in distant ages, enabling us to read by their aid the lesson of the earth's history and its mutations.

The eminent chemist, Bischof, affords us explanation on the general conservation of organic remains. He says:

"The cause of the destruction of animal remains enclosed in sandstone is easily explained. These rocks are porous.

from careful examination, ascertained the character and form of these tiny creatures. It had hitherto been supposed that no living organism existed in any deep-ocean bed, and but for the aid of novel and specially adapted mechanical appliances we should have continued to remain ignorant of the knowledge of such jelly-like vitalism, not only maintaining life, but capable of constructing their skeletons out of the carbonate of lime contained in the water in which they lived; these so-called Globigerina, when dead, becoming imbedded in a granular matrix composed of various bodies so minute in character that a cubic inch represents hundreds of thousands compacted together by innumerable granules.

The chalk is not solely composed of Globigerina, but contains a small percentage of siliceous bodies of a low type of vegetable organism named Diatomaceæ, together with the simple animals termed Radiolariæ, both of which live and die on the surface of the ocean. The time occupied by these siliceous organisms, scarcely equalling the lightest dust in weight, in gravitating to their final resting place in the ocean bed, must have been very protracted, more especially when we consider the great disparity between their superficial area and weight.

If we assume that during the geologic periods of the marine limestone formations the amount of carbonic acid was as great as that now existing, we find ample confirmation that such deposits were the result of organic agency.

Bischof tells us that in our present seas there is seven times as much carbonic acid gas as is necessary to hold in solution the carbonate of lime, and that therefore its deposition on the sea bed could only be accomplished by vital forces.

From an analysis made of the water flowing off from the mud of the river Hooghly, in the delta of the Ganges, after the annual inundation, it was found to be highly charged with carbonic acid gas holding lime in solution.

That the minute agencies contributing and collectively bringing about such results were various in character, is apparent from the fact that, with but slight exceptions, each formation has its distinct representative fossils. So also do we find that these remains indicate the particular epoch in which they lived and died.

We have shown that these vast masses of extinct organisms were accumulated in ancient seas. Such a conclusion finds ample proof from the existence, in a fossilized state, of numerous remains of aquatic animals who passed their lives in the sea, the majority of them being similar to those now found in existing oceans.

A good illustration of the ceaseless activity of submarine life is afforded by the following information:

"A remarkable piece of coral has been taken from a submarine cable, near Port Darwin, Australia. It is of the ordinary species, about 5 inches in height, 6 inches in diameter at the top, and about 2 inches at the base. It is perfectly formed, and the face bears the distinct impression of the cable, while a few fibres of the coir rope, used as a sheath for the telegraph wire, still adhere to it. As the cable has been laid only four years, it is evident that this specimen must have grown to its present height in that time; this seems to prove that the growth of coral has been much more rapid than scientific men have hitherto admitted."

We seldom find chalk in so pure a state on the earth's surface as that found in the now forming ocean beds, various geological changes having occurred since that remote period, and those of more recent date especially have left ample evidence of their occurrence by deposits of a purely diluvial character. It is from the intermixture of these impurities that the cement maker is to apprehend danger, from the consequences of which he can only protect himself by the exercise of the most careful attention.

The next deposit we purpose considering is the lias, which we regard as second in importance—from the cement-making point of view—to that which we have just referred to.

Lias.

This formation is so named, it is supposed, from a corruption of the word lyas, or layers, and was originally applied to the thin-bedded limestones at the base of the oolitic system. It is generally arranged in the following divisions:

1st. Upper lias clay, or shale.
2nd. Marlstone.
3rd. Lower lias clay, or shale.
4th. Lias rock.

Although, perhaps, not equalling the chalk in geographical extent, it, however, occupies a considerable portion of this country's surface, and extends in almost unbroken continuity from Lyme Regis in the south, to the river Tees in the north, with isolated patches in North and South Wales and elsewhere. In Scotland and Ireland it exists but in comparatively insignificant quantities. The most favourable points for studying and observing this highly interesting geological formation are Whitby, in Yorkshire; Westbury, in the estuary of the river Severn; Watchet, in Somersetshire; Aberthaw, in Glamorganshire; Lyme Regis, in Dorsetshire; and Barrow, in Leicestershire.

Besides its industrial value this formation commands much attention from its being the depository of the gigantic fossil remains of the ichthyosauri, plesiosauri, and other extinct animals. Its prevailing and distinguishing colour is blue, its shade varying in different districts, even reaching that of a pale grey or white, known in some localities as the "white lias," when it assumes a highly indurated character, and when in that state is used for ornamental and other constructive purposes.

This formation, from its muddy appearance, and the almost entire absence of any crystalline characteristics, induces a belief that some different natural process was at work than that which built up the chalk and other carbonate of lime deposits. The disproportion of the limestone to the shale is very marked, and although the former is seldom found with a larger proportion than 80 per cent. of carbonate, and then in comparatively shallow beds, the clays preponderate, and even in the best circumstanced quarries, unless where brickmaking is carried on, large quantities are wheeled to waste, thereby involving an increase in the cost of working.

The perfect animal remains found in the "lias" induces the belief that the growth of its deposit was at a rapid rate, the unconsumed food, and other conditions of the exhumed remains, favouring this supposition. That the deposit was performed in still and placid waters is beyond question, from the accuracy—even to the colour in the shells—of the most delicate remains. The ooze bed of the great liassic sea formed a soft resting place for the delicate pentacrinites, as well as the huge saurians, both of whom, according to their weight, became entombed, to be recovered in distant ages, enabling us to read by their aid the lesson of the earth's history and its mutations.

The eminent chemist, Bischof, affords us explanation on the general conservation of organic remains. He says:

"The cause of the destruction of animal remains enclosed in sandstone is easily explained. These rocks are porous.

They are therefore saturated by the sea water, and, after their upheaval by the ocean, by rain water; the salts of lime of which these remains consist, are carried away.

"If dead marine animals sink into sandbanks, they can, of course, only be enclosed by sand where currents exist; for sand at the bottom of the sea can only be pushed forward, never lifted, by the water. Where no new deposits of sand took place the bones of the animal remains were exposed to the dissolving action of the carbonic acid contained in the sea water, after their flesh had become decomposed. Even if the body has been covered with sand at the bottom of the sea, and this sand has been cemented together, yet the bones are not protected from the dissolving action of sea water; for no sandstone excludes water.

"It is quite a different case with those substances held in suspension by the water. These belong to the whole ocean. They sink to the bottom whenever the water is not kept in continual motion by storms. Dead marine animals are there enveloped by these substances and thus preserved. They are also preserved if enveloped in carbonate of lime, whether this has been produced by the decomposition of the bicarbonate, by the atmosphere, or by organic action.

"Although clay-slate, clay, and limestone may not be completely impervious to water, yet they are more so than sandstone. The circumstance is also favourable that water penetrating limestone layers becomes saturated with carbonate of lime before reaching the organic remains, and can therefore hardly act upon them. Thus we find an abundance of remains of fish and saurians in the bituminous schists or limestone of the upper lias in several countries. The schists in which the remains of the ichthyosauri—those voracious monsters—occur, show everywhere such a uniform and remarkable composition, that one is inclined to ascribe the formation of those layers to the animals which are enclosed in them. Under the magnifying glass you can see that they are composed of fishes' teeth, scales, fragments of shell, &c., and all is impregnated with the animal oil which renders these schists combustible. . . . We need not look for a mechanical power to account for the grinding up of the fishes' teeth, &c. We find one in the digestive apparatus of the saurians which fed on the fish."

In the absence of such authoritative hypothesis we might feel disposed to look around for other natural agencies assisting in the formation of this remarkable geological deposit. We have instances innumerable of remarkable results from apparently trivial causes, such, for instance, as that stated by Agassiz occurring in the river Glat—one of the tributaries of Lake Zurich. Thousands of barbels were destroyed by a sudden reduction of 15° in the temperature of the water. A sudden irruption of salt water into a fresh-water lake resulting in a similar disaster.

Volcanic agency is another active instrument of destruction. A comparatively recent instance of this power occurred at Quito in 1797. At the foot of the mountain of Tunguragua a stream of water and fetid mud poured out, overflowed, and carried destruction along their course. Valleys 1000 feet broad were flooded to the depth of 600 feet, and the course of the rivers completely choked by the deposited mud; the vapour killing all the cattle coming within its poisonous influence.

In the Crimea there are active mud volcanoes, and in the neighbouring peninsula of Tamsau an outbreak of Kuknobo in 1794 emitted about 27,000,000 cubic feet of mud.

The initial force by which some or all of these streams were propelled was in all probability derived from the elastic power of steam under enormous pressure. This agent, when relieved of the incumbent equilibrium by which previously its latent energy had been controlled, discharging the mud, and thereby preventing, it is possible, destructive earthquakes.

A singular phenomenon occurred in France during the recent heavy rains which fell during twenty-nine days over Paris. It was computed that during that time the quantity of water which fell amounted to 4,500,000 tons (about one thousand million gallons). From an analysis made it was estimated that about nine tons of ammonia had been precipitated from the atmosphere, in which was contained nitrogen sufficient to cover the whole of Paris with a forest. Besides the ammonia there were 88 tons of mineral substances, amongst them being globules of meteoric iron.

It was from the lias deposits that Smeaton selected the now famous Aberthaw lime, with which he mixed Italian puzzolana, for the mortar used in the construction of the Eddystone Lighthouse.

The analyses of Aberthaw limestone are—

No. 1.

Carbonate of lime	86·2
Silica and alumina (clay)	11·2
Water, &c.	2·6
	100·0

No. 2.

Carbonate of lime	86·0
,, magnesia	2·0
Silica	8·0
Alumina	1·0
Peroxide of iron	2·0
Water and loss	1·0
	100·0

The following analyses of lias limestone and shale are taken from Rogers' 'Iron Metallurgy,' and made presumably by the author thereof, or under his control and direction. The various specimens were submitted to him for investigation to ascertain their value in relation to iron making.

		Carbonate of Lime.	Carbonate of Magnesia.	Silica.	Alumina.	Peroxide of Iron.	Water and Loss.
1	Black lias, Mumbles, near Swansea	50	20	4	2	20	4
2	Flesh-coloured ditto, ditto	37	34	5	2	14	8
3	Yellow ditto, ditto	50	28	4	2	8	8
4	Lime shale, Llangattock, Brecounshire	34	4	24	2	16	20
5	Upper ditto, ditto	38	4	20	4	12	22
6	Yellow coarse ditto, ditto	75	2	10	2	4	7
7	Lias stone, ditto	48	30	8	4	4	6
8	Reddish ditto, ditto	80	8	4	2	4	2
9	Blue sandy ditto, ditto	30	..	64	2	2	2
10	Blue smooth ditto, ditto	58	22	8	4	4	4
11	Lias, near Bridgend, Glamorganshire	70	6	16	2	4	2
12	Ditto No. 2, ditto	76	2	12	2	6	2
13	Ditto No. 3, ditto	76	4	10	2	4	4

There are other well-known districts from which hydraulic limes have been obtained and used in important engineering works. From Lyme Regis the mortar used in the construction of the London Docks was obtained; and in the no less important undertaking of the Liverpool Docks, that from Holywell, in North Wales.

Their analyses are as under:

	Lyme Regis.	Holywell.
Carbonate of lime	79·20	71·55
Alumina and silica	17·30	23·60
Water	3·50	0·50
Carbonate of magnesia	..	1·35
Oxide of iron	..	2·21
Alkalies	..	0·79
	100·00	100·00

An exceptionally good specimen of the Warwickshire lias limestone, analyzed by W. F. Reid, gives:

Carbonate of lime	68·25
Silica	17·90
Alumina	3·70
Oxide of iron	2·01
Magnesia	1·44
Sesquioxide of manganese	0·14
Organic substances	0·50
Water and loss	6·06
	100·00

Lias limestone, from Larne, in Ireland, analyzed by Dr. Hodges, gives:

Carbonate of lime	71·66
Carbonate of magnesia	2·67
Oxides of iron and alumina	9·42
Silica and insoluble clay	14·61
	98·36

The singularly varied character of the several beds of this formation prevents our being able to add to the above limited number of analyses. In fact, every band of limestone and shale has its own distinct and separate chemical value. The stone may be regarded as worth in carbonate of lime from 60 per cent. to 80 per cent., and the shales seldom contain more than 30 per cent., or less than 10 per cent. In some districts it is not easy to discriminate between the newly quarried stones and shales. After a short exposure to the air, however, shales disintegrate, and in that condition are readily distinguished from the stone. Like almost all other shales, those from the lias beds are found mixed with iron pyrites. As this substance (sulphuret of iron), when existing in certain proportions, exercises an injurious influence on the manufacture of Portland cement, great care is necessary in the selection of shales for that purpose. It is probably owing to the presence of this deleterious ingredient that so little progress has been made in producing *first-class* Portland cements from these deposits; and the ease with which all the required materials are obtained may also have conduced to a careless system, disregarding the necessity of being guided by chemical knowledge.

The favour with which engineers generally regard hydraulic limes encourages its manufacture, to the exclusion of cement, notwithstanding that for all constructive purposes it is much more costly and less valuable in its results for mortar-making purposes than a good Portland cement. The manufacture of the ordinary lias lime is a simple A B C operation, and requires neither the outlay of large capital nor much technical knowledge.

Geologists have given more than ordinary attention to this deposit, and the late Hugh Miller, in writing of that at Eathie, in Scotland, says:

" The liassic deposit of Eathie must have been of slow deposition. It consists of laminæ as thin as sheets of pasteboard, which of course shows that there was but little deposited at a time, and pauses between each deposit, and, though a soft muddy surface could have been of itself no proper habitat for the sedentary animals—serpulæ, oysters, gryphites, and terebratulæ—we find further, that they did, notwithstanding, find footing upon it, by attaching themselves to the dead shells of such of the sailing or swimming molluscs, ammonites, and belemnites, as died over it, and left upon it their remains; from which we infer that the pauses must have been very protracted, seeing that they gave sufficient time for the terebratulæ—shells that never moved from the place in which they were originally fixed—to grow up to maturity. The thin leaves of these liassic volumes must have been slowly formed and deliberately written; for, as a series of volumes, reclining against a granite pedestal in the geologic library of nature, I used to find pleasure in regarding them. The limestone bands, curiously marbled with lignite, ichthyolite, and shell, formed the stiff boarding; and the thin pasteboard-like laminæ between — tens and hundreds of thousands in number in even the slimmer volumes—composed the closely written leaves. For never did figures or characters lie closer in a page than the organisms on the surfaces of these leaf-like laminæ. Permit me to present to you from my note-book a few readings taken during a single visit from these strange pages.

"We insinuate our lever into a fissure of the shale, and turn up a portion of one of the laminæ, whose surface had last seen the light when existing as part of the bottom of the old liassic sea, when more than half the formation had still to be deposited. Is it not one of the prints of Sowerby's 'Mineral Conchology' that has opened up to us? Nay, the shells lie too thickly for that, and there are too many repetitions of organisms of the same species. The drawing, too, is finer, and the shading seems produced rather by such a degree of relief in the figures as may be seen in those of an embossed card, than by any of lighter or darker colour. And yet the general tone of the colouring, though dimmed by the action of untold centuries, is still very striking. The ground of the tablets is of a deep black, while the colours stand out in various shades, from opaque to silvery white, and from silvery white to deep grey. *There*, for instance, is a group of large ammonites, as if drawn in white chalk; *there*, a cluster of minute bivalves, resembling pectens, each of which bears its thin film of silvery nacre; *there*, a gracefully formed lima, in deep neutral tint; while lying athwart the page, like the dark hawthorn leaf in Bewick's well-known vignette, there are two slim sword-shaped leaves coloured in deep umber. We lay open a portion of another page. The centre is occupied by a large myacites, still bearing a warm tint of yellowish brown, and which must have had an exceedingly brilliant shell in its day; there is a modiola, a smaller shell, but similar in tint, though not quite so bright, lying a few inches away, with an assemblage of dark grey gryphites of considerable size on the one side, and on the other a fleet of minute terebratulæ, that had been borne down and covered by some fresh deposit from above, when riding at their anchors. We turn over yet another page. It is occupied exclusively by ammonites of various sizes, but all of one species, as if a whole argosy, old and young, convoyés and convoyed, had been wrecked at once, and sent disabled and dead to the bottom. And here we open yet another page more. It bears a set of extremely slender belemnites. They lie along and athwart, and in every possible angle, like a heap of boarding pikes thrown carelessly down on a vessel's deck on the surrender of the crew. Here, too, is an assemblage of bright black plates, that shine like pieces of japan work, the cerebral plates of some fish of the ganoid order; and here an immense accumulation of glittering scales of a circular form. We apply the microscope and find every little interstice in the page covered with organisms. And leaf after leaf, for tens and hundreds of feet together, repeat the same strange story. The great Alexandrian library, with its unsummed tomes of ancient literature, the accumulation of long ages, was but a poor and meagre collection, scarce less puny in bulk than recent in date, when compared with this vast and wondrous library of the Scotch lias.

"Now this Eathie deposit is a crowded burying ground, greatly more charged with remains of the dead, and more thoroughly saturated with what was once animal matter, than ever yet was city burying ground in its most unsanitary state."

Sir Charles Lyell, in remarking on the line in Byron,

"The dust we tread was once alive,"

says:

"How faint an idea does this exclamation of the poet convey of the real wonders of nature! for here we discover proofs that the calcareous and siliceous dust of which hills are composed has not only been once alive, but almost every particle, albeit invisible to the naked eye, still retains the organic structure which, at periods of time incalculably remote, was impressed upon it by the powers of life."

In looking around us, and applying our intelligence, we can no longer doubt or hesitate to believe in the existence of an unceasing "great cause" shaping and operating at its will on every speck of the boundless space of the universe. Volcanic action! atmospheric action! meteoric action! with innumerable minor influences, all impenetrable to the ordinary mind, but occasionally, and at rare intervals, expounded through the acute intellects of some advanced scientific intelligence. We can read, with a certain amount of clearness, the lessons taught by the volcanic outbursts of past and present times. Their external influence as it affects the physical conditions of ourselves and the surface of the earth; but the cause lies hidden, and may still be regarded as a sealed book, notwithstanding the vaunted explanation of egotistic philosophers. That the prime agent is chemical we may safely assume, even with our comparatively limited knowledge of its vast agency, and those phenomena which, during the benighted ages, had their origin ascribed to "dark" or "infernal" agency, can be readily explained by modern scientists. Take, for instance, a natural occurrence of the simplest kind, which recently took place in the adit level of a lead mine in the state of Missouri, North America. A sudden rise of the temperature from 60° to 100° obliged the miners to retire from their work. Chemistry at once read the—at least what would have been a hundred years ago—riddle, and proved that the presence of 75 per cent. of sulphate of protoxide of iron in the strata through which the adit was driven had become disintegrated, and in this condition rapidly absorbed the oxygen of the atmospheric air, resulting in the increase of the 40° of temperature.

Many a mysterious underground fire could have been similarly explained if the necessary intelligence could have been commanded.

The phenomena of metallic snow, and other kindred occurrences, should now no longer terrify the ignorant, for science has loudly proclaimed the source and cause of their occurrence.

We shall now proceed to examine the

OOLITIC FORMATION,

which, like the cretaceous and lias deposits, is well represented in England, extending from Yorkshire in the north to Dorsetshire in the south, and having an irregular breadth of about thirty miles.

The name by which geologists have distinguished the formation originated in its peculiar physical character, which resembles the roes or eggs of fish, around which are concentrically accumulated layers of calcareous matter. It is arranged by geologists into three divisions, viz.:

"Lower,"
"Middle,"
"Upper."

From each of these divisions are quarried various stones for building and lime burning. The well-known and valuable Portland and Bath stone, as well as the Purbeck marbles, fullers' earth and firestone, are the most marked products from this formation. Its use for Portland and other cement-making purposes has hitherto been but limited.

The external aspect of the rock, and its general character, is not so well-defined, and is exceedingly variable, being neither crystalline nor amorphous. Its colour is also unreliable, and exhibits almost every gradation of shade between white and yellow, according to the extent of metallic influence with which it is surrounded.

The clays which are intercalated in some of the series are generally of a dark blue colour, tolerably pure and well suited for cement making in conjunction with its allied stone.

The chemical results are somewhat varied in character, and exhibit a wide range in the various widely separated districts from which the samples have been obtained.

We here give six analyses:

      No. 1, from Portland (Dorset).
      „ 2, „ Ancaster (Lincoln).
      „ 3, „ Ham Hill (Somerset).
      „ 4, „ Barnack (Northampton).
      „ 5, „ Cirencester (Wiltshire).
      „ 6, „   „

The two latter were analyzed by Dr. A. Voelcker, with a view to an ascertainment of their value to the farmer for agricultural purposes.

	1.	2.	3.	4.	5.	6.
Carbonate of lime	95·16	93·59	79·30	93·40	95·35	89·20
„ magnesia	1·20	2·90	5·20	3·80	0·74	0·34
Iron and alumina	0·50	0·80	8·30	1·30	1·42	4·14
Silica	1·20	..	4·70	..	2·44	6·83
Sulphate of lime	..	..	..	..	0·20	0·09
Phosphoric acid	..	..	..	..	0·13	0·06
Water and loss	1·94	2·71	2·50	1·50	..	..

One valuable peculiarity in connection with this formation is the fact of its being favourably circumstanced with regard to means of transit, and in its northern and midland divisions within the influence of supplies of cheap fuel. Except in the southern fields it is far distant from the sea, and does not, like the chalk and lias, command the advantages of that means of cheap conveyance.

These three formations, viz. the chalk, lias, and oolite, have to a greater or less extent, in this country at least, furnished the materials for Portland cement making, and we have taken them in the order of their acknowledged importance. We shall now enter on the description and discussion of other geological formations not yet sufficiently tested, but, nevertheless, possessing valuable sources from which can be derived inexhaustible supplies of the necessary cement-making ingredients. First of these in importance is the

CARBONIFEROUS (COAL-BEARING) LIMESTONE.

It is also sometimes known and recognized as the mountain limestone, from the rugged character it imparts to the scenery of the districts in which it prevails. The physical character and aspect of this interesting formation differ widely from those we have already referred to. It forms bold and prominent mountain ranges, at considerable altitudes, while the chalk, lias, and oolite formations, are seldom found at any considerable height above the sea level, and on which are to be seen the most fertile valleys and hills —from the deep soils of the midland counties to the thin ones of the southern downs: in the one case producing the rich crops of wheat, and in the other the no less valuable barleys and grasses, both alike the source of agricultural wealth. The mountain limestone exhibits no such pleasant aspects, for its texture is too hard, resisting the climatal and meteoric actions which so readily degrade the other rocks to a condition fitted for agricultural treatment, their high and exposed situation also unfitting them for successful cultivation, although at moderate altitudes the grass grown is of good quality, and is much superior to that grown in the gritstone soils at similar elevations.

Generally speaking this formation is of great depth, and in some of its localities of deposit comes into close contact with the coal measures. It attains its greatest purity in the counties of Derby, Denbigh, and Flint, and in some of the Derbyshire quarries it is nearly of a pure white colour, having but an imperceptible trace of oxide of iron. This freedom from the usually associated impurities of crystalline limestones commands for it a high position with chemical manufacturers, and it is extensively used in the production of chemicals in Lancashire, Yorkshire, and Northumberland. Its purity leads to an enormous consumption at long distances from its locality of deposit.

For cement making this property is not so important, as when impurities exist they generally favour the cement maker, at all events when they exist in clearly defined and easily estimated proportions.

Owing to its generally isolated and unapproachable position the carboniferous limestone is in the more important deposits beyond the reach of the necessary silica and alumina, as well as unfavourably circumstanced in regard to a supply of fuel. There are, however, fringing on its margin, in some districts considerable deposits of aluminous and other shales from which can be readily obtained the necessary cement-making ingredients; more especially is this the case in some parts of North Derbyshire, where it comes in contact with the well-known millstone grit formation.

The following analyses of the carboniferous limestone exhibit the chemical values of the best-known deposits, viz. :

English and Welsh.

No. 1.—Derbyshire (Muspratt).
  „ 2.—Yorkshire (Johnston).
  „ 3.—Flintshire.
  „ 4.—Weardale, Yorkshire.
  „ 5.—Durham (Johnston).
  „ 6.—Cumberland (Johnston).

	1.	2.	3.	4.	5.	6.
Carbonate of lime	97·13	94·56	89·75	93·72	95·06	94·86
Sulphate of lime	..	0·32	..	..	..	0·23
Phosphate of lime	..	0·33	..	..	..	..
Carbonate of magnesia	0·18	2·32	..	3·63	2·46	1·26
Silica	0·69	1·29	0·60	0·76	1·32	2·92
Alumina and iron	4·00	1·15	8·80	1·06	1·00	0·73
Water	..	..	0·25	0·09	..	..
Bitumen	..	..	0·60	..	trace	trace
Alkalies	0·16	..	..	..	..	..
Silicic acid	0·10	..	..	..	..	..
Sulphuric acid	0·24	..	..	..	..	..

Scotch.

No. 1.—From the Coal Measures, Burdie House, Edinburgh.
Nos. 2, 3, and 4.—Slate rocks, Ardgone, Argyleshire.
No. 5.—Relig, Inverness.
  „ 6.—Cantyre.
  „ 7.—Cantyre.
  „ 8.—Fifeshire (Professor Johnston).

	1.	2.	3.	4.	5.	6.	7.	8.
Carbonate of lime	80·52	89·99	90·14	89·15	93·82	98·05	90·96	66·0
„ magnesia	0·91	5·20	0·31	2·56	1·64	0·44	0·62	9·45
Oxide of iron and alumina	5·87	0·96	0·51	0·51	0·99	0·29	1·81	8·73
Inorganic siliceous material	5·56	3·72	9·08	7·48	3·55	1·27	6·40	..
Silica	..	..	..	..	..	..	..	13·0
Water	..	..	..	..	..	..	..	1·90
Bitumen	..	..	..	..	..	..	..	0·94

Irish.

Analyzed by Dr. Hodges.

No. 1.—Magheraw, County Antrim.
  „ 2.—Glendarra, „   „
  „ 3.—Moira,   „ Down.
  „ 4.—Castle Espie, „   „

	1.	2.	3.	4.
Carbonate of lime	98·63	95·03	96·80	94·40
,, magnesia	0·38	0·55	0·76	1·38
Phosphate of lime	0·10	0·18	0·12	0·05
Oxide of iron and alumina	0·08	2·00	0·40	0·40
Silica and insoluble clay	0·45	1·20	0·56	2·40

From Brown's Cliff, Carlow.

Analyzed by Mr. Griffiths.

Carbonate of lime	95·0
Silica and insoluble clay	4·5

Dublin (Calp).

By Mr. Knox.

Carbonate of lime	68·0
Oxide of iron and alumina	9·5
Insoluble clay and silica	18·0
Organic matter	4·5

Analyzed by Mr. Jones.

No. 1.—Clones, County Monaghan.
,, 2.—Newton Gore, County Cavan.
,, 3.—Belturbet, ,, ,,

	1.	2.	3.
Carbonate of lime	89·08	65·10	98·00
,, magnesia	1·97	1·40	1·28
Oxide of iron and alumina	0·66	0·66	0·30
Silica and insoluble clay	8·16	32·85	0·42

Murlough Bay, Donegal.

By Apjohn.

Carbonate of lime	97·14
,, magnesia	2·32
Silica and insoluble clay	0·98

We have shown by the above selection of reliable analyses —obtained from widely separated localities—that in its larger deposits this limestone is comparatively pure, but in those which are isolated and limited there is frequent and considerable admixture of impurities, of such a character, however, as to add to rather than detract from their cement-making value.

In this formation are found the now famous caves on the contents and form of which so much has been said by geologists. The exploration of them made from time to time has revealed the existence of human relics commingled with the remains of extinct animals. These interesting discoveries prove that man was their occupant not simultaneously with the animals, but at intervals of time—proved by reliable scientific evidence—of great duration. This fortunate preservation of these remains has been due to the gradual deposition of the carbonate of lime—in a state of solution—gradually percolating through the fissures of the rock, and in its deposition forming stalactite on the roof and stalagmite on the floor or pavement of the caves. These caves it may be supposed were originally formed by the gradual action of a turbulent and highly agitated sea, probably intensified by steam pressure, at a period long anterior to their occupancy by man or his contemporaries.

By a careful study of these remains interesting evidence is obtained of the so-called antediluvian animals, their habits, their food, and organization. Such information is, however, of a problematical character when compared with the direct results obtained from an investigation of the contents of these caves. It is impossible to measure with any degree of accuracy the period at which man first sheltered within their rugged space. Recent explorations of "Kent's Cavern" were made under the direction of Mr. Pengelly, who, speaking in 1872, says as follows:

"Coming to the question of time (meaning the age of the several deposits in which the bones were found), we have gone back ten thousand years at least—that is the minimum, it may be more—before we get through the *black mould;* we enter then the granular stalagmite, and we know from the nature of the case that that thickness of stalagmite must indicate an enormous length of time, inasmuch as the stalagmitic floor cannot be formed faster than the limestone is dissolved overhead, and the solution of that limestone is due to the presence of carbonic acid, and there is no possibility, under existing conditions, of any other water entering that cavern than what falls on the hills as *rain*. I do not ask you to take the thickness of the stalagmite as a chronometer, but will tell you a fact. There is in one part of the cavern a huge boss of stalagmite rising up from the floor. That boss betokens that its formation was comparatively very rapid. Take that rapid rate as the measure. There is on the boss an inscription, 'Robert Hedges, of Ireland, February 20, 1688.'

"For a hundred and eighty-four years the drip has been going on, and it has failed to obliterate that inscription; the film of stalagmite which has accreted on it is not more than the twentieth of an inch in thickness. Nearly two hundred years for the twentieth of an inch, and you have five feet to account for! But whatever may have been the time necessary for the formation of the stalagmite, the cave earth is older still. There is another and more ancient stalagmite thicker still; below that there is another deposit older than all, and in that we found human implements."

Anxious lest his statements should be considered incredible, Mr. Pengelly adds this by way of confirmation, that he was in possession of McEnery's MS. (a previous explorer of the same cavern) in 1825, who described the above inscription at that date, and adds:

"The inscription then is good for forty-seven years. Further, it was not newly cut then, for he (McEnery) says the letters are glazed over with a film of stalagmite; in short, his description applies accurately to it now. We have therefore the best possible reason for believing that it is genuine in every sense."

How very modern and even recent appears " Uriconium," with its interesting records of Roman occupancy, in comparison with the undiscovered or untold age of " Kent's Cavern," or even the period at which it gave shelter to its first human tenant.

Thus by the discriminating accuracy and observation of the skilled geologist, fact upon fact proves that this " Kent's Cavern " — with others of a kindred character widely separated from each other—had been occupied by a variety of animals—Roman pottery indicating that even historic man found shelter therein. The bronze spear heads and polished stone celts affirming also the tenancy of the prehistoric or neolithic man, while the flint implements and the fossil bones of the great quadrupeds prove the contemporaneous existence of palæolithic man.

During the intervals between one period and the other there were doubtless varied climates, ranging probably between those of a torrid and almost glacial character. The numerous distinct animals created and adapted for such changeful conditions must also have been of divers characters and constitutions. There are found in these caves in confused and intermingled masses fossilized bones of the hyena, bison, reindeer, brown, grizzly, and great cave bears, and man.

Lowest in the series of limestones, and of doubtful value to the cement maker, occurs the

MAGNESIAN OR DOLOMITE LIMESTONE,

of considerable extent in this country, having its greatest range in a comparatively narrow strip between Sunderland and Nottingham. It is sometimes also called the " conglomerate limestone." The physical peculiarities by which it is readily distinguished from any of the other varieties of limestone consist in its texture and colour; the former being granular and the latter yellow. Its chemical analysis is somewhat fluctuating.

The stone used in the construction of York Minster was obtained from this formation; as also that used in building Westminster Hall. The analysis of the former, according to Mr. Smithson Tennant, is:

Carbonic acid	47·00
Lime	33·24
Magnesia	19·36
Iron and clay	0·40
	100·00

While that at Westminster Hall is about 2 per cent. less magnesia.

The following analyses are:

No. 1.—From Denton, near the Tees: by the Rev. J. Holmes.
Nos. 2 and 3.—From Eldon and Aycliffe: by Sir Humphry Davy.

	1.	2.	3.
Carbonate of lime	63·00	52·0	45·9
,, magnesia	30·00	45·2	44·6
Alumina, red oxide of iron, and bitumen	2·25		
Water	0·25		
Iron		1·1	1·6
Residuum		1·1	2·8

In the neighbourhood of Bristol Dr. Gilby found a limestone containing a considerable quantity of carbonate of magnesia, resembling in its physical characteristics that from Northumberland, which on analysis proved to be as follows:

No. 1.—From near the village of Portishead.
" 2.—From a point four miles north-west of Bristol.

	1.	2.
Carbonate of lime	53·5	58·0
" magnesia	37·5	38·0
Oxide of iron	0·8	1·1
Insoluble matter	7·0	..
Loss	1·2	1·4
Silica and bituminous matter	..	1·5
	100·0	100·0

Lime made from this stone is fairly hydraulic, and is generally appreciated for building purposes. For agricultural use it is unsuitable, owing to the injurious influence of the magnesia when carelessly or ignorantly applied. It requires great care in burning, especially when bitumen is present in the analysis.

The commission appointed to report upon the most suitable stones for building the Houses of Parliament, selected four specimens from this formation, and submitted them for analysis to Professors Daniell and Wheatstone, who reported as follows:

No. 1.—Bolsover, Derbyshire.
" 2.—Huddleston, Yorkshire.
" 3.—Roach Abbey, "
" 4.—Park Nook, "

	1.	2.	3.	4.
Carbonate of lime	51·5	54·19	57·5	55·7
" magnesia	40·2	41·37	39·4	41·6
Iron and alumina	1·8	0·70	0·3	0·4
Silica	3·6	2·53	0·8	..
Water and loss	3·3	1·61	1·6	2·3

Owing probably to the great abundance of other well-known and easily accessible carbonates of lime, the magnesian limestone has not yet received much attention from builders.

Scotch.

The following analyses are by Professor Johnston, from specimens taken at Langton, Berwickshire:

	1.	2.	3.	4.
Carbonate of lime	43·64	47·00	39·00	43·81
,, magnesia	33·49	38·04	30·25	39·53
Oxides of iron and alumina	1·06	1·99	1·39	3·57
Insoluble siliceous matter	21·50	12·97	29·27	13·09

Irish.

A sample from Holywood, County Antrim, analyzed by Dr. Hodges, gives:

Carbonate of lime	46·33
,, magnesia	44·11
Phosphate of lime	0·31
Oxide of iron and alumina	2·25
Silica and insoluble clay	5·00

Other localities by Sir Robert Kane:

No. 1.—Kilkenny.
,, 2.—County Down.
,, 3.—Dublin.
,, 4.—Sligo.

	1.	2.	3.	4.
Lime	30·13	30·26	30·2	30·3
Magnesia	21·43	18·25	20·6	22·1
Oxides of iron and manganese	0·95	3·10	1·5	0·6
Silica	5·74		1·5	
Carbonic acid	46·65	47·26	46·2	47·0

In Germany and France much attention has been directed to the manufacture of a cement from magnesite (hydrous carbonate of magnesia), obtained from Frankenstein in Silesia. The mode of manufacture is as follows:

The stone is broken into pieces about the size of the closed fist, and then burned in retort furnaces, such as are used in gasworks. After being burnt, it is ground by edge runners, sifted fine, and at the same time mixed with a certain proportion of amorphous silica.

It is used for castings, but cannot compete with plaster of Paris. When aqueous solutions of chlorides are added it is capable of attaining a high degree of induration. It is named "Albolith cement," and is used with advantage in coating plaster of Paris castings, which, when so treated, become extremely hard.

As a wood preservative it is likely to become useful. Experiments have been made in saturating railway sleepers, but as a prolonged period is necessary to prove its value for such a purpose, we must wait some time yet for the result. This cement is unable to resist the action of water, and in consequence offers no advantage to the engineer in the construction of hydraulic works. When setting it develops a large amount of heat, approaching almost to the boiling point, and consequently when used in glue moulds requires special treatment and care. This objection, however, applies only to large castings, for when used for small ones the heat is very inconsiderable.

The above allusion to this new cement is made to show that many materials, of which we know and care but little, are sometimes, by chemical ingenuity, converted into useful and profitable products. We may perhaps think that there is no possibility of getting a better cement than the "Portland," but the makers and users of Roman cement firmly and persistently opposed the introduction of that cement until its merits overcame at last all prejudices and opposition.

Before leaving the carbonate of lime division, we will shortly refer to some of the Staffordshire stones.

In South Staffordshire there are three groups of limestone

rocks, each having a variety of beds of stone, separated by shaly partings, and belonging to the Upper Silurian formation.

They are classed in three divisions as follows:

1st. Ludlow, locally termed *Sedgley stone*, or Aymestry limestone.

2nd. Wenlock rocks, producing the *Dudley and Walsall stone*.

3rd. Woolhope, from which is obtained the barn, or *Hay Head limestone*.

The analyses of these stones are as follows:

No. 1.—Aymestry or Sedgley.
" 2.—Dudley Castle stone.
" 3.—Wren's Nest stone.

	1.	2.	3.
Carbonate of lime	81·10	90·09	74·64
" magnesia	2·27	1·26	2·51
Carbonate and protoxide of iron	2·15	Alumina and Iron	1·63
Alumina	0·45	2·30	0·75
Phosphoric acid	0·21	0·46	0·35
Insoluble matter	13·38	5·13	20·03
Loss	0·44	0·76	0·09

The two following are fairly hydraulic in character:

No. 1.—*Walsall*.

Moisture	0·40
Organic matter	5·60
Carbonate of lime	76·40
" magnesia	2·10
Oxide of iron	2·50
Silica	13·00

No. 2.—*Rushall*.

Carbonate of lime	63·98
Alumina	1·00
Protoxide and peroxide of iron	5·17
Phosphoric acid, &c.	1·70
Magnesia and alkalies	1·41
Insoluble earthy matter	25·66
Loss	1·08

In the same beds are found a finer quality of limestone used as a flux in the manufacture of pig iron; two samples of which give the following analyses:

	1.	2.
Carbonate of lime	95·38	93·17
,, magnesia	0·31	..
Phosphate of magnesia and phosphoric acid	0·39	..
Phosphate of lime, iron, manganese, and alumian	..	3·03
Alumina	0·40	..
Protoxide of iron	1·37	..
Insoluble earthy matter	1·80	3·70
Alkalies and loss	0·35	..
Magnesia and alkalies	..	trace

The Woolhope limestones, locally known as,

    No. 1.—Captain stone,
    ,, 2.—Middle stone,
    ,, 3.—Bottom stone,

have analyses as under:

	1.	2.	3.
Insoluble matter	22·58	26·35	17·90
Oxides of iron	2·35	1·21	1·55
Alumina	2·30	1·70	2·13
Phosphate of lime and phosphoric acid	0·90	1·54	0·77
Carbonate of lime	70·00	67·38	75·85
Magnesia and alkalies	1·21	1·03	1·40
Moisture and loss	0·46	0·79	0·40

The insoluble matter in these three specimens on analysis gave:

    Alumina and oxides of iron   ..   ..   ..   15·30
    Silica acid and silica ..   ..   77·10
    Magnesia, alkalies, and trace of lime   ..   7·60

We are again indebted to Rogers' 'Iron Metallurgy'

for the following analyses of limestones in various formations:

		Carbonate of Lime.	Carbonate of Magnesia.	Silica.	Alumina.	Peroxide of Iron.	Water and Loss.
	ENGLAND.						
1	Brownish-white limestone, Worcestershire	88	1	4	2	3	2
2	White ditto, ditto	92	0	3	2	1	2
3	Westcott stone, Somersetshire	92	1	2	2	0	3
4	Lype ditto, ditto	94	1	2	1	0	2
5	Pool ditto, ditto	92	0	4	2	1	1
5A	Red marly stone, near Ross, Herefordshire	72	0	16	10	2	0
6	White limestone, Newport, Monmouth	94	0	2	1	0	3
7	Grey limestone, Bewdley Forest	82	0	6	2	8	2
	SOUTH WALES.						
8	Brown limestone, Llangattock, Breconshire	75	3	5	2	14	1
9	Yellow ditto, ditto	80	2	4	1	2	1
10	Dark blue ditto, ditto	72	6	10	2	10	0
11	Brown Clydach, ditto	77	6	4	2	10	1
12	Ditto, paler, ditto	70	10	5	3	12	0
13	Yellow No. 2, ditto	60	8	14	4	14	0
14	Ditto No. 3, ditto	60	7	16	4	11	2
15	White, ditto	95	0	2	1	1	1
16	Ditto No. 2, ditto	98	0	1	0	0	1
17	Ditto No. 3, ditto	98	0	1	0	0	1
18	Green, with shells, ditto	90	2	6	1	1	0
19	Dark brown, ditto	64	0	6	4	10	16
20	Dark blue, Glamorganshire	70	2	8	3	17	0
21	Ditto No. 2, ditto	50	6	18	6	18	2
22	Pink-coloured, Pentmoels, Glamorganshire	44	16	12	10	16	2
23	White crystalline, ditto	94	2	3	0	0	1
24	Rough sample ditto, ditto	96	0	2	1	0	1
25	Rose-coloured, Pyle, Glamorganshire	94	1	1	0	2	2
26	Ditto No. 2, ditto	88	2	6	2	1	1
27	White, ditto	84	5	3	2	4	2
28	Bluish white, Glamorganshire	95	2	2	0	1	0
29	Red, Pentmoel, ditto	52	18	4	2	20	4
30	Blue, ditto	67	6	10	8	5	4
31	White Midkin, ditto	93	1	2	1	0	3
32	Red Machen, ditto	64	10	2	4	12	8
33	Fawn-coloured, ditto	70	16	1	0	8	5
34	Blue, ditto	64	14	8	4	6	4
35	White, Bridgend, ditto	98	0	1	0	0	1

GEOLOGICAL AND MINERALOGICAL OBSERVATIONS.

		Carbonate of Lime.	Carbonate of Magnesia.	Silica.	Alumina.	Peroxide of Iron.	Water and Loss.
	SOUTH WALES—*continued*.						
36	White, Bridgend, No. 2, do.	97	0	1	1	0	1
37	Yellow Rhudry, ditto	72	0	6	2	14	6
38	Black, near Swansea	50	20	20	2	6	2
39	Fawn-coloured, Varteg, Monmouthshire	50	44	1	0	4	1
40	Ditto No. 2, ditto	50	46	1	0	3	0
41	Ditto No. 3, ditto	46	50	1	0	2	1
42	Ditto No. 4, ditto	50	46	1	0	2	1
43	Grey, Swansea Valley	82	2	10	1	5	0
44	Ditto No. 2, ditto	94	2	2	0	1	1
45	Ditto No. 3, ditto	76	0	15	2	4	3
46	White, ditto	98	0	0	1	0	1
47	Blue, ditto	76	10	6	2	4	2
48	Red Risca, Monmouthshire	78	12	2	0	6	2
49	Ditto No. 2, ditto	78	12	0	1	8	1
50	White, Llangattock, Breconshire	96	0	1	1	0	2
51	Yellow, ditto	86	0	4	2	6	2
52	Brown, ditto	84	6	3	1	4	2
53	Blackish, ditto	76	2	12	4	4	2
54	Fawn colour, ditto	72	22	2	0	4	0
55	Greenish, ditto	94	0	4	0	0	2
56	White, with shells, ditto	96	0	2	0	0	2
57	Fawn colour, Pwlt Caer, ditto	68	28	2	1	0	1
58	White, ditto	97	1	1	0	0	1
59	Black, Cardiff Wharf	90	2	2	2	0	4
60	Blue, Breconshire	64	12	10	5	4	5
61	Pink, ditto	56	16	14	4	1	4
62	Yellow, Aberdare Valley	62	12	6	4	14	2
63	Nodular limestone, Builth	66	2	24	2	4	2
64	Crystals of ditto, ditto	68	0	14	2	8	8
65	Nodular, Park Farm, ditto	72	0	12	2	8	6
	NORTH WALES.						
66	Blue (no locality)	84	6	8	1	1	0
67	Brown coarse (ditto)	40	6	26	1	25	2
68	Ditto No. 2.	36	4	26	4	28	2
	IRELAND.						
69	White, from Cork	98	0	1	0	0	1
70	Ditto No. 1, ditto	99	0	0	0	0	1
71	Ditto No. 2, ditto	99	0	0	0	0	1
72	Ditto No. 4, shelly, ditto	96	0	2	1	0	1
73	Grey, Limerick	96	0	2	0	1	1

These limestones are valuable, both as hydraulic limes, and for smelting purposes.

We will now proceed to consider other materials of a less

clearly defined character, which in many cases only require chemical investigation to render them useful ingredients for cement making.

CHALK MARLS

Are to be found in great abundance, and some deposits could by simple treatment be converted into good cements. Such a sample, taken from the lower stratum of the chalk formation, offers favourable advantages for that purpose. In general, however, the analysis of chalk marls is fluctuating, and therefore requires careful observation.

Carbonate of lime	66·69
Silica, insoluble	19·64 ⎫ 26·09
" soluble	6·45 ⎭
Oxide of iron and alumina	3·04
Magnesia	0·68
Phosphoric acid	1·82

Such as have been used for hydraulic limes or Portland cement have mostly exhibited the dangerous property of shrinking during setting. So that it would be safer to treat these marls under a cement-making process, and add the necessary ingredient which may be found to be deficient.

It is unsafe to deal with any of the other marls, unless when found clear of sand and clayey in character. In such condition they may be easily and economically dealt with either under the wet or dry process.

The following favourable analyses are from Rhœtic beds in the neighbourhood of Bristol.

No. 1.—Variegated Keuper marl.
 " 2.—Black marly shale.

Specimens dried:

	1.	2.
Carbonate of lime	3·87	8·96
Silicate of alumina	92·97	89·01
Silica, &c.	3·16	2·03

Some of the fresh-water or infusorial deposits of carbonate of lime, resemble in many respects some of the marls. They are not of frequent occurrence in this country, but are abundant in Italy and other parts of Europe, as well as North and South America. Below are two analyses of a deposit from Hanover, No. 1 upper, and No. 2 lower:

	1.	2.
Silica	87·859	74·48
Alumina	0·133	..
Oxide of iron	0·731	0·39
Carbonate of lime	0·750	0·34
Organic substance	2·279	} 24·43
Water	8·431	

The above show a considerable difference in the value of separate parts of the same deposit.

The following specimen of this class of deposit was submitted to the author to test its value and capabilities for Portland cement making. It is from Italy, and an analysis by W. F. Reid gave the following results:

	Unburnt.	Burnt.
Silica	1·40	2·57
Alumina and oxide of iron	0·65	1·20
Lime	52·17	95·42
Magnesia	0·46	0·81
Carbonic acid	41·40	..
Water	1·91	..
Loss and organic substance	2·00	..

This material is of low specific gravity, and contained fragments of shells in their original state, showing it to be therefore a comparatively recent formation. Some vegetable fibres were interspersed throughout the mass, which probably had been the remains of plants from the surface under which it was dug.

In conjunction with the above material, and for the same

object, a sample of marly clay of a pale yellow colour (almost white) was also submitted to the author, and by the same analyst was found to be as follows:

	Unburnt.	Burnt.
Silica	39·04	49·47
Alumina and oxide of iron	16·86	21·37
Lime	19·56	24·80
Magnesia	3·42	4·31
Carbonic acid	14·09	
Water	5·57	
Loss and organic substance	1·46	

specific gravity also light; colour when burnt, pale red. The powder clinkered at a light red heat, on account of the existence of so much lime. Before burning this clay exhibited considerable effervescence when treated with acids.

By an accurate amalgamation of these two materials the author succeeded in making a first-class Portland cement. He submitted the materials to both processes, and by each, with little difficulty, good and satisfactory results were obtained. These materials are singularly well adapted for the manufacture of a good cement at comparatively low cost.

In the various deposits of limestone we have described there are to be found innumerable varieties of fossils, a study of which would well repay the cement maker from the geological point of view, for in that direction much information is to be obtained of a useful kind.

There are deposits from which *natural* Portland cements are obtained, such as that at Boulogne, where large quantities are made from one of the layers of the Kimmeridge clay about 160 feet below the strata in which the "septaria," or "Boulogne pebbles" are found. The deposit is argillo-calcareous, and in its natural state is first burnt and then ground for use.

The analysis of this cement is as follows:

Lime	65·13
Magnesia	0·53
Silica	20·42
Alumina and small quantity of oxide of iron	13·87
Sulphate of lime	a trace

The cement is slow setting in quality, and some experimenters attribute to it the objectionable, if not dangerous, quality of shrinkage during setting.

There is some doubt as to the reliability of this cement, and it cannot at all events be equal in quality to an accurately prepared Portland cement.

At the Vienna Exhibition there was exhibited a good Portland cement, produced from the two stones, of the following analysis.

It was an exhibit from Hungary.

	1.	2.
Silica	33·10	24·18
Alumina	1·40	12·10
Oxide of iron	4·80	3·78
Carbonate of lime	44·42	54·68
Magnesia	3·20	0·71
Alkalies	4·10	0·45
Sulphuric acid		0·28
Carbonic acid	5·98	0·78
Water	0·40	0·06

We shall now proceed to the discussion of the next most important ingredient in cement making, viz.:

SILICA.

We have endeavoured to show how abundant are the supplies of carbonate of lime and their almost universal distribution; but even their abundance is eclipsed by the deposits of silica which are to be found in nearly every geological formation. They exist in much variety of form, from the precious stones of the jeweller to the building stones of the mason. Silica imparts its special characteristic to the granites, and its combination with potash, alumina, and magnesia produce mica and felspar. Basalt, serpentine, lava, syenite, are all silicates, and its presence in soils adds to their fertility and value, while in the useful art of glass making it is a leading and indispensable ingredient.

It has hitherto been supposed that the deposition and formation of siliceous rocks was solely due to the degradations and wear of other formations, but recent explorations of the 'Challenger' expedition are likely to throw some light on the existence of other vital agencies of incessant activity busy in the lowest depths of the oceans. The more familiar physical agencies, such as the *meteoric, chemical, igneous, aqueous,* may be regarded as the palpable force of the surface, while in the hidden depths of boundless seas the tiny converters of the raw material carry on their ceaseless energies hidden from our observation, and until recently unknown.

The emerald consists of 68 per cent. of silica and 15 per cent. of alumina, and the presence of from 1 per cent. to 2 per cent. of the oxide of chromium imparts to it, according to the extent thereof, its beautiful shades of green. The opal has from 90 per cent. to 95 per cent. of hydrous silica, and from 5 per. cent. to 10 per cent. of water; its variable and changing hues being due to the minute traces of peroxide of iron, potash, soda, lime, alumina, &c.

Major Ross, author of 'Pyrology, or Fire Chemistry,' informs the author of the following interesting fact.

"I have just discovered that phosphoric acid, to the amount of 1 or 2 per cent., is the *cause* of opalescence in the *opal,* a problem hitherto unsolved."

Klaproth's analyses of sapphire, catseye, emerald, and the pale bluish green Siberian beryl, are as follows:

	Sapphire.	Catseye.	Emerald.	Beryl.
Alumina	98·50	1·75	15·75	16·25
Oxide of iron	1·00	0·25	1·00	0·60
Lime	0·50	1·50	0·25	..
Silica	..	95·00	68·50	66·45
Glucine	..	..	12·50	15·50
Oxide of chromium	..	..	0·30	..
Loss	..	1·50	..	..

Amethysts, cairngorms, agates, bloodstones, carnelians, catseyes, onyx, &c., are only so many varieties of rock crystal, the water of which contains the metallic oxides which impart the varied tints and colours; and we may regard them all as specimens of glass from nature's laboratory and the produce of her unerring handiwork.

The finest crystals are found in the mountains of Switzerland, Ceylon, Madagascar, and the Brazils, the latter producing the pebbles for spectacles and the glasses of optical instruments.

In calling attention to the silica and alumina division of our subject, we will give precedence to the clays which are almost invariably used with the chalk, and in combination with which produce in the easiest and cheapest manner Portland cement.

Alluvial Clays.

The term "alluvial" is not strictly accurate, as in its geological sense it is more appropriately applied to the deposits of "alluvium" of a mixed and heterogeneous character. It would be more correct therefore to term this deposit one of a water or marine character, and of comparatively modern origin.

The following observations on the Thames and Medway

deposits are, with the exception of local differences, equally applicable to other tidal and river agency in nearly all parts of the world.

The so-called alluvial clays are best known to cement makers from their local association with the chalks, and from having always formed one of the materials used by early experimenters in their search for hydraulic cements, those of the Thames and Medway more especially obtaining almost a world-wide reputation, and indeed until the application of chemical knowledge were believed to be the only clays by and through which Portland cement could be made.

Good selected samples of these clays have been subjected to many analyses, but we will confine ourselves to the following, viz. No. 1 by Feichtenger, No. 2 Faraday, and No. 3 Recent.

	1.	2.	3.
Silica	68·45	64·72	70·56
Alumina	11·64	24·27	14·52
Lime	0·75	1·89	4·43
Oxide of iron	14·80	7·14	3·06
Soda and kali	..	..	..
Soda	2·90 } 4·00	..	3·95
Potash	1·10 }		
Carbonic acid	..	..	3·48

Owing in some measure to the favourable source of its origin from the wasting and degradation of the cliffs and shores of the Isle of Sheppy, and Kent, and Essex, it is found abundantly, and in certain localities of unexceptionable purity. These cliffs and shores are much exposed to climatic and marine action, and owing to their being situated for the most part on the London clay formation, readily succumb to such degrading action. The clay so detached is readily taken up by the water, and the more soluble parts conveyed to considerable distances, according to the state and velocity of the tides.

At the time when the Medway clay was first used as a cement-making agent, there were extensive tracts of saltings or marshes in the estuaries of the rivers Thames and Medway formed by this sea action. These marshes when embanked or reclaimed became in that protected state valuable grazing land, in high reputation for cattle rearing. Around the coasts there are many similar reclamations, the most important of which occurs in the counties of Cambridge and Lincoln, by which an estuarine bay, seventy miles long and upwards of thirty in breadth—formed from the denudation of the Kimmeridge and Oxford clays—by engineering skill has become perhaps one of the most fertile districts of this country.

This fine river clay became an object of considerable value to the cement maker, from the facility with which it can be amalgamated with chalk, and continues to be used by the cement makers whose works are situated on the banks of the rivers Thames and Medway. In the condition in which it is excavated there is a large amount of moisture, estimated by Pasley in his experiments to be $\frac{11}{20}$, or more than one-half of its gross weight. The deposits are of varied character, consequent on the condition of the water during their period of deposit; in calm weather during neap tides, securing the best conditions for the deposition of the finer particles, and during spring tides and high winds, the coarser parts of the wasted shores were precipitated from their waters of solution. The use of this clay may be ascribed to the fortunate circumstance of Pasley having selected Chatham as his point of experiments. This was in 1828, and it is probable that Aspdin—whose first patent was dated in 1824—in his remote field of practical investigation, remained ignorant of Pasley's discovery.

Deposits of pit and other clays are to be found in various conditions of purity in almost every district of this and other countries, from the finest porcelain clay to the common kinds of brick earth.

The purest porcelain clay (Kaolin), according to Forchhammer's analysis, is

Silica	47·0
Alumina	39·2
Water	13·7

while Phillips gives it as follows:

Silica	71·15
Alumina	15·86
Lime	1·92
Water	6·73

Other examinations differ in value, as shown by the following analyses from specimens obtained at Mosl, near Halle. No. 1 (washed) by Forchhammer, No. 2 (raw) by Stephens, No. 3 (raw) by Michaelis.

	1.	2.	3.
Silica	46·80	67·58	68·371
Alumina	36·83	22·67	17·976
Oxide of iron			
Lime	3·11		
Magnesia		0·46	0·202
Carbonic acid			
Lime	0·55		
Potash	0·27		
Water	12·44	7·85	8·300

A sample from Cornwall has

Alumina	60 parts.
Silica	40 ,,

Dorsetshire Clays.

Nos. 1 and 2, and No. 3 undried.

ANALYSES.

	1.	2.	3.
Silica	65·49	72·33	68·95
Alumina	21·28	23·25	22·25
Iron oxide	1·26	2·54	
Alkaline earths	7·25	1·78	
Sulphate of lime	4·72	trace	
Water			8·80

Potters' clay is of variable quality, and requires to be used with much care. Its chemical value ranges thus:

Silica	from 44 per cent. to 58 per cent.
Alumina	„ 24 „ to 38 „
Iron oxide	„ 1 „ to 7 „
Water	„ 10 „ to 15 „

White Clay, near Bristol.

Silica	63·99
Alumina	22·71
Water	13·30
	100·00

Again we have—especially in the "coal measures"—abundance of fire-clays of high cement-making value. Those in Staffordshire, and Stourbridge in Worcestershire, are the best known, from their having been long used in the manufacture of bricks for iron making, and other purposes.

The following analyses exhibit their silica-alumina value:

No. 1.—White fire-clay, Tamworth.
„ 2.—Black clay, „
„ 3.—Burnt clay, Stourbridge.
„ 4.—Clay, Tintern Abbey.
„ 5.—Clay, Amblecote.
„ 6.—Best Glasshouse clay, Tamworth.

	1.	2.	3.	4.	5.	6.
Silica	59·87	57·45	67·69	72·75	70·32	75·99
Alumina	33·49	37·93	27·91	22·37	26·42	22·53
Protoxide of iron	3·01	2·20	2·35	2·18	1·04	0·97
Lime	1·42	0·92	0·63	0·58	0·36	0·05
Magnesia	0·31	0·15	0·11	0·22	0·43	..
Soda and potash	trace	trace	..	..	1·40	0·87

Analyses of the celebrated Dinas fire-clay by Dr. Percy:

	1.	2.
Silica	98·31	96·73
Alumina	0·72	1·39
Protoxide of iron	0·18	0·48
Lime	0·22	0·19
Potass and soda	0·14	0·20
Water combined	0·35	0·50

Analyses of an artificial plastic so-called "Dinas-crystal," made by Messrs. Reith and O'Brien at Bonn. No. 1 analyzed by Dr. Carl Bischof, and No. 2 by Dr. H. Seger.

	1.	2.
Silica	87·48	87·89
Alumina	4·66	7·17
Oxide of iron	2·62	0·82
Lime	1·08	0·95
Carbonic acid, water, &c.	3·96	3·04

The above is a yellowish-grey mass containing grains of quartz. It neither shrinks nor expands in the fire; but does not resist such high temperatures as the genuine Dinas brick.

Clay from the river Mersey, near Liverpool, gives:

Silica	37·75
Alumina	21·58
Carbonate of lime	} 40·67
Sulphate of lime and oxide of iron	

This clay has been used for cement making in conjunction with London chalk.

It is evident from the above varieties of clays on the sea and river shores, or in other deposits, whether formed by the precipitation of degraded rocks or through volcanic or organic agency, that abundance—if not unlimited quantities—of clays can be commanded by the cement maker.

A valuable cement-making clay can be obtained from the deposit known as the "gault." The analysis gives:

Silica	46·61
Alumina	16·06
Carbonate of lime	24·95
Oxide of iron	6·07
Alkalies	6·18

Scotch Fire-clays.

No. 1.—Garnkirk, near Glasgow.
 „ 2.—Gartcosh, „ „

	1.	2.
Silica	53·4	60·96
Alumina	43·6	37·00
Protoxide of iron	1·8	1·16
Lime	0·6	0·64
Magnesia		0·24
Moisture	0·6	

In the selection of clays from this group, great care should be bestowed on the examination of their mechanical or physical character, and preference given to those, where a choice is possible, of most perfect homogeneity, and capable of the readiest conversion into a state of plasticity.

For blending with amorphous carbonate of lime (chalk) the alluvial clays are very suitable, and during the first forty years of the "Portland cement" period commanded the sole attention of the manufacturer. Their capacity of easy and simple conversion to the required purpose led to a hap-hazard system of manufacture, and a total disregard of the chemical element in the process.

To show what care is exercised in the selection of clays for cement making on the Continent, we give the analyses of clays selected and analyzed by Dr. Michaelis.

No. 1.—From the province of Saxony, on the Elbe.
" 2.—From Pomerania.
" 3.—From the Upper Hartz.
" 4.—From Brandenburg.

	1.	2.	3.	4.
Silica	60·06	59·25	60·00	62·48
Alumina	17·79	23·12	22·22	20·00
Oxide of iron	7·08	8·53	8·99	7·33
Lime	9·92		4·18	6·30
Magnesia	1·89	2·80	1·60	1·16
Potash	2·50	1·87	1·49	1·74
Soda	0·73	1·60	0·72	0·37

These clays resemble those of the Medway in their analyses, but are found in a more economical form in a dry state.

We shall now proceed to consider other more clearly defined sources from which silica and alumina can be obtained, with the advantage of being more reliable in character and free from the inconveniences attending the use of river clays.

There can be no doubt, as we here admit, of the extremely favourable circumstances which surround the manufacturer of Portland cement, in the so-called "London district," and the cement makers so happily situated have no desire to alter the stereotyped character of their established processes. They are, however, limited to the use of the chalks and clays, for within their reach (with the exception of flints) there do not exist other materials for their purpose, but if there were, their adoption would necessarily involve an undesirable change in the economy of their works.

Flints furnish a valuable auxiliary to the cement maker by their being almost pure silica. Klaproth gives the analysis as,

Silica	98·00
Lime	0·50
Alumina	0·25
Oxide of iron	0·25
Water	1·00

The specific gravity being, according to Brisson, 2·594.

Flints are of various shades of colour, from grey to black, and are found sometimes hard enough to scratch quartz. It is infusible when in a state of purity, but becomes white at a moderate temperature. The most abundant deposits are found in the upper bed of the chalk formation. On exposure to air and water they become yellow, and in that condition are known as ferruginous flints, such as are found in gravel beds, which by the action of water have been rounded by attrition. It is difficult, if not impossible, to grind flints in their natural state fine enough for amalgamation with limestone or chalk, but when they have been exposed to the temperature of a dull red heat they are easily ground to a fine dust or powder.

A cement has been prepared, called "Chalcedony cement," from flints, by H. Frühling, at Bliesshastel. In a description of the process he says:

"Flints of the right kind are a good substitute for dear cements, when they occur in sufficient quantity and where the machinery of a cement manufactory is at hand for their conversion."

He prefers fresh-dug flints from their original bed of deposit, and when in that state reduced to a fine powder, act similarly to puzzolana. This powder, when mixed in the following proportions, produces a good hydraulic mortar.

One part of slaked lime, a quarter part of powdered flints, and three parts of sand.

Flints containing a large percentage of lime melt at a high temperature into a vesicular and semi-transparent slag. The right temperature for roasting the flints is produced by a mixture of from 9 to 10 parts of flints to 1 of coke, and the operation performed in a common kiln. It is, however, preferable to use a kiln with a regenerator, as the flints need not in that case be so finely broken and the temperature can be regulated by the *burner*.

After burning, crushing, and grinding, the product is a light bluish grey angular powder, partly soluble in a solution of caustic potash, and giving a gelatinous paste when immersed in concentrated muriatic acid. Exposure to the atmosphere has an injurious effect on the powder, and it should, therefore, to avoid deterioration, be packed in casks.

The best way to use the powder is to mix it with equal parts of rich lime, to which add from 2 to 3 parts of sand. The mortar thus prepared binds or sets very slowly, and resembles in character those mortars made with rich limes and puzzolanas. After from three to five days it resists the action of water, and gradually hardens until ultimately it excels all similar preparations. Mixtures of burnt flint or chalcedony cement, mixed with lime in the proper proportions, make a good Portland cement, but at a cost too high to permit of such application where siliceous clays or marls are cheaply accessible.

However, in one direction, this cement appears to be most valuable, and from its colour could be used with advantage in wall decorations, either in exposed or sheltered situations. For this purpose (for which it appears to be invaluable) it is used in the following proportions: viz.—1 part of lime paste; 1 volume of flint cement; and 2 parts of white sand.

In these proportions it produces a white shining plaster, which becomes extremely hard. The trowel used for polishing its surface should be made of copper.

The main object of this division of our treatise is to point out the existence of such materials as possess in a high degree the necessary elements of successful cement making, and obtainable in districts where the cost of London cement is practically prohibitory of its use.

When a proper and intelligent acquaintance with the chemistry of our subject has been established, the engineer or architect can, without any fear or misapprehension, secure the cement for his work at almost any point where a sufficient quantity may be required to warrant its local manufacture for special purposes.

Having thus directed attention to the mud or clay deposits associated with the chalks, we will next consider the valuable deposits of silica and alumina of the lias formation, so fortunately allied and intermingled with abundance of carbonate of lime.

In the comparatively limited quarries of the lias districts widely scattered in remotely separated counties, we are unable to judge accurately of their extent or value. Until recently they were used only as sources for hydraulic lime, in the manufacture of which the valuable shales were regarded as waste. In Warwickshire the excavations have been carried on at depths limited only by the cost of quarrying and the water to be kept clear of the quarrymen; but in Leicestershire—more especially the Barrow district—the excavations are limited in depth, and when the required rock has been obtained, the ground is levelled and again resumes its original agricultural aspect. In Warwickshire, more especially, much attention is now given to the manufacture of Portland cement, and many of the hitherto rejected beds of stone and shale are becoming more valuable, and instead of being buried at considerable cost are utilized for cement purposes. There is but a small percentage of these deposits suitable for simple lime making, and that is only found near the surface. There have been some attempts to make a natural cement from one of the beds, but owing to an excess of carbonate of lime, there is too much risk and danger attending it to warrant its continuance. The disproportion of the shales to the stones, amounting generally in the proportion of from three to six of the former to one of the latter, entails an expense which makes the cost of the stone when saddled with the rejected shales a heavy item in cement making, in these otherwise favourably situated districts. It should be seriously considered by the proprietors of these quarries whether it is politic to send off the hydraulic lime in such quantities as to render the conversion of the shales into Portland cement comparatively impossible.

For the reasons before given it is difficult to give reliable analyses of these shales, but the two following of lias clays in Somersetshire will give some idea of their value.

No. 1.—Green lower lias.
" 2.—Lower lias clay.

	1.	2.
Silica	34·50	33·75
Alumina	29·05	17·92
Sulphate and carbonate of lime, and oxide of iron	17·20	48·33
Water	19·35	
	100·00	100·00

The majority of the lias shales contain a considerable portion of crystals of iron pyrites, so much so indeed in some beds as would in favourable conditions of the sulphur market pay for its extraction, and leave the shale pure and in the best condition for cement making. The external characteristics of these shales when freshly quarried are most variable; but it is found by experience that those of fine and even texture are the most suitable for cement purposes. It would be safest to deal with lias shales direct, and have the chemical value of each bed accurately ascertained, for in the heterogeneous condition in which they are found when promiscuously heaped up in the spoil banks they would be a source of trouble and anxiety in consequence of the diversity of their analyses.

In the more irregularly and less defined deposits of Somersetshire, Worcestershire, Yorkshire, and North and South Wales, there is much less, and indeed in some instances an entire absence of the shales or clays. They have, however, an advantage over the inland deposits of having the command of sea transit, and of being generally within the reach of an economical supply of fuel.

Oolitic clays are not so abundant or so pure, although in

the deposits of Northamptonshire clays of a reliable character are of easy obtainment. An exceedingly favourable section of this formation is commanded by the London and North-Western Railway cutting between the stations of Blisworth and Roade. If not so prolific in shales, there are abundant beds of silica of varied character intercallated between the beds of stone, which would serve as good substitutes for the clays.

In the carboniferous or mountain limestone formation there is a scarcity of clays, and except where it comes in contact with the gritstone, the supplies of silica and alumina are uncertain and limited in extent. In the North Derbyshire division there are considerable deposits of basalt, known locally as toadstone, dunstone, &c., and in the north-western margin a fringe of shales of considerable depth. In both of these localities abundance of silica is to be obtained for admixture with the limestone. The beds of shale have at various depths layers of septaria from which might be produced a good Roman cement. The shales, however, are of varied character, generally impregnated with iron pyrites and sulphur, and therefore to be cautiously used for cement making. The following analyses are from the shales near Buxton:

No. 1.—Disintegrated shale.
Nos. 2 and 3.—Hard shales.

Analyzed by W. F. Reid.

	1.	2.	3.
Silica	49·78	50·67	57·00
Alumina and oxide of iron	29·07	28·37	24·33
Magnesia	1·11	1·98	2·10
Manganese	1·43	trace	..
Lime	0·40	4·14	1·88
Water	12·84	12·35	12·27
Alkalies and loss	5·37	..	..
Alkalies and sulphur	..	2·49	..
Iron pyrites	..	..	2·36

With these shales in conjunction with lime waste and limestone—under the author's patent—a large quantity of Portland cement is now being manufactured by the dry process. The site of the works being favourable for a cheap supply of fuel and within the easy reach of good markets for the sale of their produce.

The author has also made an excellent cement with the toadstone and mountain limestone. The analysis of the former by W. F. Reid is as follows:

Silica	43·59
Alumina	18·46
Oxide of iron	10·92
Lime	10·61
Magnesia	6·39
Water	2·21
Carbonic acid	4·41
Alkalies	3·31

The presence of carbonate of lime indicates a partial decomposition of the sample, which was in a slightly disintegrated state.

Analysis of Derbyshire toadstone by Dr. Withering:

Specific gravity	2·3
Silica	63
Alumina	14
Carbonate of lime	7
Oxide of iron	16
	100

The toadstone is a basalt, having a somewhat earthy fracture, and when exposed on the surface becomes pulverulent, forming a fertile soil whereon timber grows luxuriantly, contrasting agreeably with the otherwise sterile limestone tracts, in which they may be regarded as the oases. The presence of intrusive basalt in this district is interesting not from its novelty, for it is found similarly placed in other parts and in Ireland and Scotland, but from the peculiar manner in which it is imbedded between the limestone deposits. It seldom exhibits itself on the surface, at least rarely in comparison with the extent of the country it traverses. The beds vary in character, some being basaltic and compact, others again earthy, vesicular, and amygdaloidal. The distinguishing and most general term toadstone in Derbyshire is called in German "todt stein," or dead stone, such designation being applied to its non-metalliferous character, for lead veins running through the limestone are cut off at their contact with this erupted rock; the term toadstone of the Derbyshire miner originating doubtless in the resemblance to the spots on the back of the toad, more especially when the rock partakes of an amygdaloidal (almond-shaped) character.* There are various names applied to this basalt, such as black stone, dunstone, channel, black clay, &c., but they are of purely local significance and without scientific value.

The lead mining operations carried on from a remote time in this part of Derbyshire, and at an early period of geological knowledge offered favourable opportunities for the study of this basaltic rock and its influence on the surrounding rocks. An early geological observer, Mr. Whitehurst, in writing on this toadstone in 1792, observed as follows:

"A blackish substance, very hard; contains bladder holes, like the *scoriæ* of metals or Iceland lava, and has the same chemical property of resisting acids. Some of its bladder holes are filled with spar, others only in part, and others again are quite empty. This *stratum* is not laminated, but consists of one entire solid mass, and breaks alike in all directions. It does not produce any minerals or figured stones representing any part of the animal or vegetable creation, nor any adventitious bodies enveloped in it; but is as much an uniform mass as any vitrified substance whatever; neither does it universally prevail as the limestone *strata;* nor is it like them equally thick; but in some instances varies in thickness from six to six hundred feet.

* It is possible, however, that the name may be an inheritance from the German miners who were employed in ancient times in this part of Derbyshire.

It is likewise attended with other circumstances which leave no room to doubt of its being as much a lava as that which flows from Hecla, Vesuvius, or Etna.

"All these circumstances plainly evince that toadstone was formed by a very different law from the others, and was greatly posterior to them; for the beds of limestone must have been formed before they were broken, and broken before their fissures were thus filled up; therefore we may, with much reason, conclude that toadstone is actual lava, and flowed from a volcano whose funnel, or shaft, did not approach the open air, but disgorged its fiery contents between the strata in all directions."

More recent and perhaps better-informed observers have arrived at a somewhat similar conclusion, and the following analyses confirm these views, at least so far as the chemical question is affected.

No. 1.—Basalt, Staffordshire, by Klaproth.
 „ 2.—Basalt from Fingal's Cave, by Streng.
 „ 3.—Basalt from the Giants' Causeway, by Streng.
 „ 4.—Toadstone, from Derbyshire (before quoted), by W. F. Reid.

	1.	2.	3.	4.
Silica	48·0	47·80	52·13	43·59
Alumina	16·0	14·80	14·87	18·46
Oxide of iron	16·0	13·08	11·40	10·92
Oxide of manganese	..	0·09	0·32	..
Lime	9·0	12·89	10·56	10·64
Potash	..	0·86	0·69	} 3·31
Soda	4·0	2·48	2·60	
Magnesia	..	6·84	6·46	6·39
Muriatic acid	1·0	..	..	..
Carbonic „	..	..	..	4·41
Water	5·0	1·41	1·19	2·21

In the neighbourhood of the Giants' Causeway the basalt rests upon the hard or indurated chalk, and is generally separated by a bed of clay of varied depth. This clay being deficient in lime and alkali is probably the result of disintegration from the parent basalt rock.

The analyses as under of this clay and the superincumbent basalt are, No. 1, clay, by Dr. Upjohn, and No. 2, basalt, by Dr. Kennedy.

	1. Clay.	2. Basalt.
Silica	50·75	48·00
Alumina	20·87	16·00
Peroxide of iron	15·90	16·00
Lime	0·72	9·00
Soda		4·00
Muriatic acid		1·00
Water	10·50	5·00
Loss	1·26	

These chemical estimates of long extinct volcanic action have their analogous representation in active modern eruptions at the Faroe Islands and Iceland.

An analysis of the former lava or basalt by Durocher gives:

Silica	46·80
Alumina	14·40
Oxide of iron	12·20
Oxide of manganese	2·80
Lime	10·16
Magnesia	9·53
Soda and potash	1·16
Water	3·00

Other lavas:

    No. 1.—Meissner, analyzed by Girard.
      „   2.—Wickenstein, Silesia, by Löewe.
      „   3.—Crouset (Haute Loire), by Ebelmen.
      „   4.—Eger, Bohemia, by Ebelmen.

	1.	2.	3.	4.
Silica	52·96	44·90	36·1	43·4
Alumina	16·46	18·71	30·5	12·2
Magnetic oxide of iron	5·32			
Protoxide of iron	2·82	9·09		12·1
Peroxide of iron			4·3	3·5
Lime	8·79	12·90	8·9	11·3
Magnesia	9·32	7·14	0·6	9·1
Soda	3·60	6·58	0·9	2·7
Potash	1·19	0·68	0·6	0·8
Water			16·9	4·4

Lava from Etna (eruption 1669), by Löewe:

Silica	48·83
Alumina	16·15
Protoxide of iron	16·32
" manganese	0·54
Lime	9·31
Magnesia	4·58
Soda	3·45
Potash	0·77

The conjunction of the basaltic or silica and alumina-bearing rocks with the crystalline chalk—before referred to—render the manufacture of Portland cement in the locality of the Giants' Causeway a matter of easy attainment.

For manufacturing purposes basalt has already been used to a considerable extent for constructive and ornamental architecture. The experiments were made from the "Rowley rag," a basaltic product obtained at the village of Rowley Regis, seven miles from Birmingham, having analysis as follows:

Silica	46·0
Alumina	16·0
Protoxide of iron	19·5
Lime	11·0
Manganese	3·5
Volatile matters	4·0

The fusibility of basalt was proved at the beginning of the century, but no attempt to utilize it was made until 1851, when Mr. Adcock patented a process for manufacturing the Rowley basalt into various useful articles. The inexpensive raw material was, however, more than counterbalanced by the high cost of the required fuel of manufacture. There were two processes adopted for the fusion of the refractory basalt, in one of which crucibles were used, placed in a reverberatory furnace, and out of which was poured the molten rock into ordinary sand moulds, previously heated and afterwards allowed to cool slowly in the ovens in which they were moulded or cast. By this process window sills, pilasters, coping, &c., were produced. The other process for the production of roofing slates was similar to that used for the manufacture of plate glass. The fused basalt was cooled until it became plastic, and in that condition placed on a metal table and rolled to the required shape and thickness. Sheets measuring 8 feet by 3 feet 8 inches were thus formed, and were capable of resisting the wearing action of the atmosphere, being lighter than slates or tiles. By this process a variety of forms could be produced, and the sheets could be cut by a diamond, like glass, its specific gravity being about the same.

This industry was carried on for nearly three years, and was then discontinued owing to its not proving a commercial success. It succumbed to the overwhelming expense of first establishment. The initiatory experiments and large cost for moulds and other appliances were too much for this infant industry. Although a financial failure, it established the practicability of converting this natural basalt into useful and graceful forms as aids to the architect. The best examples resulting from this industry are to be found in the Edgbaston Vestry Hall, built in the Anglo-Norman style, the columns, window pieces, doorways, and ornamental steps being made from the melted Rowley rag.

The molten mass, when cooled suddenly, was converted into black obsidian, or volcanic glass; but when allowed to cool slowly, resumed its original crystalline basaltic structure, without any apparent difference beyond the obliteration of the lines of cleavage.

Since the time when these successful attempts were made great strides have taken place in the knowledge and application of heat for all purposes of manufacture, and it is just possible that this improved knowledge might yet lead to a more prosperous conversion of the Rowley rag.

In the north of England, near Newcastle, occurs a basalt, known locally as the "whin sill," and closely associated with the carboniferous limestone of that district. Geologists have differed as to the mode of its formation, but it is now regarded as an intrusive rock, as originally classed by Professor Sedgwick. It appears to be similar in many respects to the Derbyshire "toadstone," and is found in horizontal *strata*, forced by the pressure by which it was originally impelled between the planes of bedding.

Here again exists, in one of the most favourably circumstanced localities, in juxtaposition, all the elements of successful cement making, favoured in a high degree with the command of cheap fuel and advantageous outlets by sea and land.

The basalts, from which are obtained the most useful materials for road and street purposes, whether in the form of broken *Macadam* or setts for paving, occur in the districts of Shropshire and North Wales—of the former from Clee Hills, and the latter ranging from Portmadoc in the west to near Conway on the east coast of North Wales. In Manchester and Liverpool these materials are used almost exclusively, unless where in the less populous districts, and where the traffic is but light and limited, the gritstone setts are considered sufficient. The excellent condition of the streets of Manchester is entirely due to the foresight and judgment of the engineer of that city, who may fairly claim the merit of raising it to the position of the best-paved town in England, and who has availed himself, with advantage, of these readily accessible materials. It may here be mentioned that after an experience of many years the larger setts have been superseded by the smallest, which are now only "$6'' \times 4''$," and prove the most suitable for the large and heavy traffic of this great manufacturing centre.

There are now several patents for the utilization of the slag from iron and other furnaces, and considerable success has been attained in making building bricks from this material. There is, however, some danger to be apprehended from the action of the weather. In Staffordshire it has been found unprofitable to macadamize the roads with the waste slag, owing to its liability to disintegration, and its use is now abandoned, Rowley rag and Welsh basalts being substituted, at a very much greater cost. In fact, the waste slag cost was but nominal, from its abundance in the iron-making districts.

In the neighbourhood of the carboniferous limestone formation, near Kendal, in Westmoreland, we find a favourable cement-making material in great abundance. It is a slate deposit, of imperfect cleavage, although tolerably compact and even in texture. Its analysis by W. F. Reid is:

Silica	60·27
Alumina	18·48
Oxide of iron	7·13
Lime	0·89
Magnesia	3·62
Oxide of manganese	5·17
Water	4·45

The river Dee, in winding through the vale of Llangollen, in North Wales, intersects at several points the limestone and slate deposits, where the manufacture of Portland cement could be performed with more than ordinary advantage, and where, owing to the moisture of the climate, it could be used for many purposes where its present cost is all but prohibitory.

An analysis of the Welsh slates is as follows, and does not differ essentially from that obtained near Kendal:

Silica	60·50
Alumina	19·70
Protoxide of iron	7·83
Lime	1·12
Magnesia	2·20
Potash	3·18
Soda	2·20
Water	3·30

Pasley, in his experiments, tried a mixture of chalk, with slate dust, but probably owing to his imperfect knowledge of the chemistry of cements (then in its infancy) he failed to produce an hydraulic cement. He tried various proportions, but found all of them unable to withstand the action of the water on being immersed. In the description of these experiments Pasley quotes an analysis of slate made by Kirwan, which was:

Silica	38
Alumina	26
Magnesia	8
Lime	4
Peroxide of iron	14

a very different result from those we have given above.

Analyses of two varieties of Slate from County Tyrone, Ireland, No. 1 being a quartz slate, and No. 2 an imperfect clay slate.

	1.	2.
Silica	67·02	52·45
Alumina	14·92	12·01
Peroxide of iron	11·36	17·92
Lime	0·73	7·72
Water	2·41	6·08
Magnesia, alkali, and loss	3·56	3·82

In Ireland there are large deposits of slate materials in many parts of the country.

A clay slate (roofing) analyzed by Abuisson.

Silica	50·01
Alumina	34·74
Protoxide of iron	3·73
Magnesia	0·87
Soda	0·04
Potash	7·21
Water	3·27

Before concluding this rather interesting part of our subject we will refer to another important source from which silica and alumina can be readily obtained. In the igneous or Plutonic formations of almost universal distribution innumerable varieties of rocks abound, closely resembling one another in chemical value.

The lavas and puzzolanas will have our first attention, from the fact of their being so well known as the agents of hydraulicity used in the preparations of the early hydraulic mortars.

The well-known and active volcanoes of Vesuvius and Etna emit abundance of silica and alumina in varied condition, which, in their natural state, on being mixed with caustic lime, render it hydraulic. There are numerous examples of the success which has attended such combinations, as illustrated by the experiments of Smeaton, Vicat, Berthier, and others.

The following are some of the analyses of this class of minerals:

No. 1.—From near Rome, by Berthier.
 „ 2.—From Vesuvius, by Vicat.
 „ 3.—From Crater of Monti Nuovo, by Abich.
 „ 4.—From Island of Vivara, by Abich.
 „ 5.—From Vesuvius during its eruption in October 1868, by Silvestri.

	1.	2.	3.	4.	5.
Silica	44·5	46·5	56·3	51·0	39·0
Alumina	15·0	10·5	15·3	13·7	14·0
Oxide of iron	12·0	29·5	7·1	13·6	13·0
Lime	8·8	10·0	1·7	7·1	18·0
Magnesia	4·7	..	1·4	4·7	3·0
Potash	1·4	..	6·5	2·9	1·0
Soda	4·1	..	8·8	2·9	10·0
Chlorine	..	..	0·3	..	..
Water	9·2	0·25	6·7	4·6	..

Santorin earths and trass are closely allied in chemical value to the above, as is shown by the following analyses:

    No. 1.—Trass, by Berthier.
      „ 2.—Trass, by Vicat.
      „ 3.—Santorin earth, by Elsner.
      „ 4.—Santorin earth, by F. Schulze.

	1.	2.	3.	4.
Silica	57·0	46·5	68·5	65·5
Alumina	16·0	20·7	13·3	16·5
Oxide of iron	5·0	5·6	5·5	3·1
Lime	2·6	2·3	2·4	2·9
Magnesia	1·0	1·0	0·7	1·5
Potash	7·0	}15·50{	3·1	4·3
Soda	1·0		4·7	2·3
Water	9·0		..	1·4
Chlorine	..	..	0·3	3·5

Again, trachyte, trachytic porphyry, phonolite, andesite, obsidian, pumice, and other kindred volcanic products, are, with slightly varying proportions—though generally exceeding them in silica value—similar to the above, and whether as unaltered mixtures with caustic lime, or more elaborately prepared combinations with carbonates of lime, they are alike deserving the consideration of the cement maker and engineer.

In some countries various volcanic deposits are the source from which are obtained blocks for building. In his recent voyage in the 'Challenger' Sir Wyville Thomson came across the following interesting evidence of the most primitive use of tufa. He says:

"The sixteen or seventeen families who reside on the island of Tristan d'Acunha, which is about half-way between the Cape of Good Hope and Cape Horn, have suffered so much from violent gales that they now build their cottages of blocks of stone—a sort of soft volcanic tufa—of four or five feet square, in order to enable them to withstand these storms. The wind was sometimes so violent that these blocks, when being brought down from the quarry, were lifted bodily by the wind. As there was no mortar on the island, all the stones were dovetailed into each other, and it was curious to see the people building these cottages. Very frequently wrecks occurred on the island, and a number of large spars were picked up on the shore as the remains of lost vessels. They got two or three of these spars, and, laying them up against the wall at a low angle, had them carefully greased, and, by a method which was known to have been used in Assyria and Ancient Egypt, they gradually moved on rollers, and slid up these blocks to the top of the wall, when they were fixed in their places. Tristan d'Acunha is one of a little group of three islands, one of which is called Nightingale Island."

The next class of rocks to which we will call attention are abundant in this country, viz. the extinct volcanic products in England, Ireland, and Scotland, in the shape of granites, traps, and basalts. Their reputation as building and paving agents is too well known to require any description in that direction at our hands, and we will therefore confine ourselves to pointing out their chemical value in relation to cement making.

Granite from a geological point of view is the primary rock on which all other formations rest, and which we may reasonably infer contributed largely in supplying some of the materials at least of all subsequent formations. The mighty and varied agencies through and by which this marvellous primeval rock was deposited, and the influences it exerted on all future geologies, are too remotely connected with our subject to require much consideration. The abundance of these rocks in Great Britain and Ireland contributes in a high degree to the picturesqueness of the locality in which they occur, and render attractive remote districts which would otherwise be uninteresting, as they are in an agricultural sense valueless.

Granite must have had its origin from sources of

great depth, and after eruption cooled and crystallized slowly under the influence of enormous pressure. Mr. Sorby in his microscopic examinations proved the existence of cavities containing fluids, from which he calculated, according to their extent or volume, the relative depths at which granite was formed in different localities. He thus estimates, that the granites of the Highlands of Scotland indicate a pressure of 26,000 feet of superincumbent rocks more than those of Cornwall. This would depend, however, on the temperature at which these rocks were consolidated. Granites vary considerably in mineralogical character, and their texture is influenced by the fluctuating qualities of felspar, quartz, mica, &c., of which they are composed. They do not exhibit much diversity in their chemical analyses, as the following selection shows:

No. 1.—Large-grained granite, from Streitberg, in Silesia.
  „ 2.—Large-grained granite, from Fox Rock, near Dublin.
  „ 3.—Fine-grained granite, from Heidelberg.
  „ 4.—Medium-grained granite, from Blackstairs Mountain, County Wexford, Ireland.
  „ 5.—Granite, from Baveno, Italy.
  „ 6.—Egyptian granite, or syenite.

	1.	2.	3.	4.	5.	6.
Silica	73·13	73·00	72·11	73·20	74·82	70·25
Alumina	12·49	13·64	15·60	13·04	16·14	16·00
Oxide of iron	2·58	2·44	1·53	1·72	1·52	2·50
Oxide of manganese	0·57		0·26			
Lime	2·40	1·84	1·26	0·96	1·68	1·16
Magnesia	0·27	0·11	0·34		0·47	
Potash	4·13	4·21	5·00	4·80	3·55	9·00
Soda	2·61	3·53	2·27	3·18	6·15	
Water	0·53		0·83			0·65
Loss		1·20				

The granites are comparatively neglected, and their valuable properties in consequence continue a blank to the artisan and manufacturer. Attention is, however, being directed to them in some directions, and a good example of how much may be accomplished by an intelligent application of scientific or technical knowledge is shown by the treatment of the raw material from which the beautiful and much-admired Beleek porcelain is made. The rock is an orthoclase granite found near the works at a point where the waters of Lough Erne flow into the river Erne. The red orthose felspar retains its crystalline form in its original beauty and perfection, and when calcined becomes white. The iron which separates itself during calcination is extracted by magnets from the powdered clay when in a moist state.

Granites from Finland, used for making bottle glass, analyzed by Struve:

	1.	2.
Silica	75·06	77·71
Titanic acid	0·36	0·48
Alumina	11·70	10·13
Peroxide of iron	1·04	1·41
Protoxide of iron	1·57	2·15
Protoxide of manganese	trace	trace
Lime	1·01	1·13
Magnesia	0·19	0·21
Potash	6·25	4·50
Soda	2·56	1·85
Water	0·63	0·44

The porphyries approximate closely in chemical resemblance to the granites, as shown by the following analyses:

No. 1.—Quartz porphyry, from the Hartz Mountains.
" 2.—Quartz porphyry (Elvanite), from County Wexford, Ireland.
" 3.—Rose-coloured quartziferous porphyry, from Grenville, Canada.

	1.	2.	3.
Silica	74·44	72·33	72·20
Alumina	13·51	8·97	12·50
Oxide of iron	2·25	7·46	3·70
Lime	1·19	1·98	0·90
Magnesia	0·01	trace	..
Potash	5·31	2·07	3·88
Soda	1·40	5·83	5·30
Water	1·34	1·86	0·60

Greenstone (Diorite, Diabase, &c.) is, like basalt, granite, &c., of igneous origin.

The following samples are from the "Sanctuaries," St. Menan, Cornwall:

	1.	2.	3.
Silica	47·66	47·32	47·79
Alumina	17·50	17·15	17·83
Oxide of iron	21·94	22·60	22·49
Lime	4·20	4·03	4·10
Magnesia	trace	trace	trace
Potash	2·43	2·33	2·15
Soda	5·19	5·27	5·88
Sulphur	trace	trace	trace
Sulphuric acid	0·16	0·18	trace
Titanic acid	trace	trace	trace
Water	0·83	0·18	0·76

Dolorite, as the origin of the name indicates, is a rock of deceptive and uncertain character.

No. 1.—From Fifeshire, Scotland, analyzed by Drysdale.
 „ 2.—From St. Austell, Cornwall, by Ebelmen.

	1.	2.
Silica	45·20	51·4
Alumina	14·40	15·8
Protoxide of iron	14·00	6·0
„ manganese		
Peroxide of iron		10·7
Lime	12·70	7·8
Magnesia	6·55	2·7
Soda	5·22	4·3
Potash		
Water	2·40	

Some years ago a cement factory was established at Radstin, near Prague, for making cement from greenstone (Diabase) in combination with primitive and transition limestones.

The analysis of the greenstone (Diabase) was:

Carbonate of lime	2·60	} Amount soluble in acid.
„ magnesia	1·00	
Oxide and protoxide of iron, with little alumina	16·30	Together, 19·90 per cent.
Silicates	79·25	

The limestones were, No. 1, primitive limestone from Cimelitz; No. 2, transition limestone from Foditz.

	1.	2.
Carbonate of lime	97·00	97·05
,, magnesia		1·41
Oxide of iron and alumina	2·00	1·40
Silica	1·00	0·22

The limestone was burnt and then slaked, with which the powdered Diabase was intimately mixed in a pasty or semi-plastic state. The facility with which the Diabase fused resulted at first in obtaining only vitrified balls. The Diabase was more finely ground, and the result was a clinkered mass of the required quality. The burning took place in Hessian crucibles in a blast furnace, after which the crucibles were allowed to cool, from which the cement was taken and ground. It was proved that a long continuous heat was more beneficial than too high a temperature. It was found that the best proportions were three parts by weight of burnt lime to two of Diabase, which was equivalent to from 30 to 32 per cent. of silicates to from 68 to 70 per cent. of lime, and the soluble compounds of the Diabase. The high percentage of iron in the raw mixture resulted in a dark-coloured cement somewhat slow in setting, but otherwise possessing the usual characteristics of a good Portland cement.

Serpentine is another rock of metamorphic origin, and may some day become an important ingredient in the economy of cement making when the exact influence of magnesia has been ascertained and valued.

The specimens from which the following analyses are made were obtained as follows:

No. 1.—Red serpentine, from Kynance Bay, Cornwall.
,, 2.—Light green serpentine, from Galway.
,, 3.—Dark green serpentine, from Col-de-Pertuis, Vosges.
,, 4.—Green homogeneous serpentine, from Oxford, Canada.

	1.	2.	3.	4.
Silica	38·29	40·12	40·83	40·30
Alumina	..	2·00	0·92	..
Protoxide of iron	13·50	3·47	7·39	7·02
Oxide of manganese	..	..	trace	0·26
,, chromium	..	..	0·68	trace
Magnesia	34·24	40·04	37·98	59·07
Lime	..	..	1·50	..
Water, &c.	12·09	13·36	10·70	13·35

Witherite, or carbonate of baryta, has been used with considerable success by Dr. Julius Aron in making a cement which in some of its characteristics and properties exceeds in value the best Portland cement. The baryta was used instead of limestone mixed with clay, obtained from the left bank of the Oder, at Stettin.

The analysis of the baryta was:

Baryta	69·29
Carbonic acid	22·95
Lime	3·85
Residue, &c.	3·20

indeed almost a pure carbonate of baryta.

The clay used had the following analysis:

Silica	54·82
Oxide of iron	6·38
Alumina	17·21
Carbonate of lime	2·39
,, magnesia	2·14
Potash	3·09
Soda	0·92
Sulphur	2·12
Carbonic acid	2·61
Water, &c.	7·58

Proportions varying from 5·75 to 4·26 parts of baryta were mixed with one part of the clay. The first mixtures, 5·75 and 5·01, after being finely ground were burnt in a sample kiln with the following results:

No. 1, 5·75 baryta and 1 clay, after being submitted to the heat of the kiln for forty-seven minutes, produced a clinker of a dark green colour. The resulting powder was grass green. When mixed with water it developed great heat, amounting to 50° C., increased in volume so much that it could only be retained in the mould by immersing it in cold water. In fifteen minutes afterwards the mould was taken out of the water, and it was found that the cement was still soft. When it hardened in the air the cement became as hard as stone, but immersed in water again it became gradually soft.

No. 2, 5·01 baryta and 1 of clay, was burnt quickly, gave a grey powder, and developed considerable warmth when mixed with water, and the mixture was of a dark bluish green and set rapidly, but became soft when immersed in water.

No. 3, 4·64 baryta and 1 of clay, burnt in forty-two minutes, gave a dark green clinker, from which was obtained a grey powder. An increase of temperature was almost inappreciable when little water was used. In five minutes the cement set so that it could leave the mould and be marked with a lead pencil. A change of colour in the cement arose, turning from an ashy grey to a dark bluish green. After several days' immersion in water the samples became soft.

No. 4, 4·26 parts of baryta and 1 of clay. The clinker of this mixture was dark green, the powder being ashy grey, and on mixing it with water an increase of temperature took place. In ten minutes the cement had set, and after four hours the nail could make only a slight impression on it. This cement also, however, when put in water softened.

For so far these experiments indicate that baryta cement possessed different qualities from that made from carbonate of lime, and was in all the proportions tried deficient in hydraulic value. The silicate of baryta, which is analogous to the silicate of lime, appears to be far more soluble in water.

Another experiment was tried with the proportions of 369 parts of baryta (witherite) to 100 of the clay. The clinker produced was dark green and the powder ashy grey. When mixed with water it set slowly, and appeared to be deficient in cement-setting properties. In the morning, however, after having been mixed sixteen hours, it attained great hardness, and had slightly loosened itself from the mould. Its tensile strength was tested, and, compared with Stettin cement, exhibited the following results.

Stettin cement, eight days old, 5 square centimetres section, broke at 86·400 kilos. = 17·280 kilos. per square centimetre; at eighteen days old, similar section, it broke at 121·350 kilos., or = 24·270 kilos. per square centimetre.

Baryta cement, with the same section, of 5 square centimetres, broke as under in sixteen hours, 135·300 kilos. = 27·060 per square centimetre; showing, therefore, that a good Portland cement (which the Stettin cement is) in eighteen days did not equal the baryta cement sixteen hours in the air. A sample of this cement, after four days' hardening in the air, broke with a tensile strain of 50·220 kilos. per square centimetre—a strength never attained from cement made of carbonate of lime.

These proportions having resulted in such satisfactory breakings, a larger quantity was made of the same mixture. The samples were made up of 214 grammes cement to 40 grammes water, which was found to be the best proportion, and again, in comparison with Stettin cement, exhibited the following results:

No. 1.

Pure Stettin Portland Cement.

Time of Setting.	Age.	Section.	Breaking Weight.	Breaking Weight per sq. cent.
15 minutes.	4 days.	5 sq. cent.	71·55 kilos.	14·31 kilos.

One-half Stettin Cement mixed with one-half Baryta Cement.

| 15 minutes. | 4 days. | 5 sq. cent. | 127·05 kilos. | 25·41 kilos. |

Pure Baryta Cement.

| 56 minutes. | 4 days. | 5 sq. cent. | 312·00 kilos. | 62·40 kilos. |

All the three samples had lain under water during the four days, and the result is good evidence that slow setting cement of any fabrication in the long run proves the best.

Further experiments were made with the baryta cement with the following results:—

	Age.	Breaking Weight per sq. centimetre.
No. 1	3½ hours.	24·36 kilos.
„ 2	11 days.	53·46 „
„ 3*	13 „	42·09 „
„ 4	23 „	53·11 „

* No. 3 sample had lain five or six days in water, but was taken out, having shown signs of cracking.

The analysis of the cement from which these experiments or tests were made was:—

Silica	17·04
Peroxide of iron	2·83
Alumina	4·88
Baryta	69·27
Lime	4·75
Magnesia	0·74

The results of these interesting experiments are given for the purpose of showing that a cement much excelling

Portland cement in its highest values can be made without chalk. It is not possible to introduce such a manufacture in this country, owing to the scarcity and value of witherite, but where great strength is required in works above water, this expensive cement might be used with advantage.

The samples were very heavy and highly indurated, so much so, as to admit of being polished like marble.

The specific gravity of the witherite, or carbonate of baryta, is $4·3$, chalk being only $2·3$, and the more compact limestones $2·6$.

Gneiss, a rock of frequent occurrence, and which usually prevails in the immediate neighbourhood of granite, for which it is sometimes from their great similarity mistaken, is to be found in almost every geological age, and is regarded as the original stratified rock, from which condition it has been metamorphosed by the action of the molten granite. To such an extent has this change in its primary structure been carried as in many instances to obliterate its original stratified character. Originating in all probability from the detritus worn from the granite in the then troubled and perhaps boiling seas, the metamorphic action by which it subsequently became changed destroyed nearly all trace of its aqueous origin. The main characteristics which distinguish it are the presence of irregularly laminated veins of striated combinations of mica, quartz, and felspar, very frequently, indeed, it may be said invariably, distorted into folds and convolutions of the most complicated character.

This rock is represented as occupying nearly one half of the north of Scotland, is frequently met with in Ireland, and Devon and Cornwall.

With this rock we will conclude our reference to the natural producing silica and alumina sources. In doing so we would call special attention to the necessity for a careful selection of these materials, guided by a previous physical and chemical examination. Some of the granites contain quartz in an unfavourable form for cement-making purposes, and by preference those of the fine-grained kind should be selected, where a choice exists.

Wherever any of these rocks are obtainable, in their disintegrated condition, they should be preferred, except when by their exposure to the atmosphere the alkalies have been washed away.

In this necessarily hurried and somewhat imperfect description of the natural depositories or storehouses accessible to the cement maker for his purposes much has been overlooked, and probably also a good deal carelessly described. We should, however, under whatever shortcomings our efforts may be surrounded, hope that a desire for further inquiry may be aroused, and if so, we feel sure that a new era will begin for the cement trade, and its valuable powers and capacities be made more generally useful.

CHAPTER III.

ARTIFICIAL SOURCES FROM WHICH SOME CEMENT-MAKING MATERIALS MAY BE OBTAINED.

Iron Slag.

WIDE as we have endeavoured to show the field of selection from which may be obtained all the necessary materials for the manufacture of Portland cement in the various geological formations, and which may be regarded as the natural supply, there are still to be found valuable supplies of an artificial character. Of those varying in importance and quality we may select as most noteworthy the slags of various kinds resulting from iron making and other allied industries, the waste lime heaps rejected by the lime burner, the gas and soap lime wastes, as well as those produced in the alkali and alum manufactures.

The abundance of the slags, lime heaps, alkali, and alum wastes in the various districts in which they occur, cannot but strike with surprise the simple observer, who must think that these industries have been of a more than usually profitable character to permit the waste of such apparently useful material. The iron and chemical trades of this country, however, have had such long runs of prosperity in times past as not only to render the manufacturers indifferent to the extent of the waste produced by their operations, but has also led to a most suicidal treatment of the minerals, as well as their extraction from their various beds of geological deposition.

A general survey of our mining districts would exhibit an amount of accumulated waste—both from coal mining and iron making operations—of a formidable character. In the coal districts the immense heaps of calcined shales indicate the existence of some arbitrary clause in the leases compelling the lessee to such a mode of destruction. In many of the more favourably circumstanced districts, where the shales are in some degree bituminous, much valuable material has thus been dissipated. These heaps are, however, now beginning to be utilized in certain districts as ballast for railways and also as a mixture for concrete building. For both purposes they are exceedingly well adapted.

We shall first consider the slags from their now in many districts attracting considerable attention, owing to their suitability for conversion to building purposes as well as with the burnt shales used by railway companies for ballasting the permanent way of their lines.

In an industry like iron making, carried on in widely separated districts, and dealing with ores and fuels of diverse character, we are sure to find the resulting wastes differing much in their chemical value and physical characteristics.

In the following analyses we have selected those best suited for illustration from a cement-making point of view, and the cognate process of slag brick or block making for building purposes.

TABLE No. 1.—*Slags from South Wales.*
(From Rogers' 'Iron Metallurgy.')

		Silica.	Lime.	Magnesia.	Oxide of Manganese.	Alumina.	Peroxide of Iron.	Excess.
1	Tawny cinder slag, Nantyglo	50	26	4	..	14	8	2
2	Black scouring ditto, Blaina	52	18	2	..	12	28	12
3	Ditto ditto, ditto	46	25	3	..	10	24	8
4	Black solid slag, Tredegar	41	34	2	2	8	23	10
5	Grey glassy ditto, Ebbw Vale	47	34	3	1	12	3	0
6	Black solid ditto, Rhymney	44	24	4	0	18	18	8
7	Ditto ditto, Dowlais	48	22	2	0	16	12	4
8	Yellow solid ditto, Cyfarthfa	42	30	2	0	14	10	8
9	Ditto ditto, Blaenavon	46	31	4	0	14	10	5
10	Black solid ditto, Nantyglo	46	28	2	0	14	12	2

From Clarence furnace, Durham, smelting Cleveland ironstone, analyzed by (1) Bell and (2) Percy, both of which show a considerable increase of lime over the above tables of South Welsh slags. Used in the Bessemer process.

	1.	2.
Silica	27·68	27·65
Alumina	22·28	24·69
Lime	40·12	40·00
Magnesia	7·27	3·55
Protoxide of iron	0·80	0·72
" manganese	0·20	0·35
Sulphur	2·00	1·95
Potash	..	0·46
Soda	..	0·99
Phosphorus	..	0·26

TABLE No. 2.—*Sundry Districts.*

(From Phillips' 'Metallurgy.')

Analyzed (Nos. 1 and 2) by Riley, (No. 3) Forbes, and (No. 4) Dr. Percy.

		Silica.	Alumina.	Lime.	Magnesia.	Protoxide of Iron.	Protoxide of Manganese.	Sulphide of Calcium.	Alkalies.
1	Grey iron slag, Dowlais	38·48	15·13	32·82	7·44	0·76	1·62	2·22	1·92
2	White iron ditto, ditto	43·07	14·85	28·92	5·87	2·53	1·37	1·90	1·84
3	Iron slag, Wednesbury, Staffordshire	39·52	15·11	32·52	3·49	2·02	2·89	2·15	1·06
4	Ditto, Dudley, Worcestershire	38·76	14·48	35·68	6·84	1·18	0·23	0·98	1·11

It will be observed that the analysis of slag produced in preparing for the Bessemer convertor is unusually favourable as a cement-making agent from the high percentage of lime it contains.

TABLE No. 3.
(From Dr. Osborn's 'Metallurgy of Iron and Steel.')

		Silica	Lime	Alumina	Magnesia	Protoxide of Iron	Manganese	Sulphide of Calcium
1	Kirkless Hall, Wigan, produced from Hematite pig-iron for the Bessemer process. In appearance this slag resembles Wedgwood ware	31·46	52·00	8·50	1·38	0·79	2·38	2·96
2	Cleveland blast furnace	32·81	34·90	23·40	7·46	0·59	0·18	1·78
3	Ditto ditto	32·94	34·94	23·32	7·33	0·46	0·09	1·88
4	Anthracite blast furnace Gartsherrie and Govan, near Glasgow	35·34	38·72	20·47	..	..	5·89	1·35

The average composition of English and Belgian slags producing foundry iron by coke hot blast is:

        Silica   .. .. 50·00
        Alumina   .. .. 23·00
        Lime   .. .. 27·00

Daunemora, Sweden, slag from Bessemer pig process. Charcoal blast.

        Silica   .. .. 46·371
        Alumina   .. ..   4·301
        Lime   .. .. 38·640
        Magnesia   .. ..   7·400
        Protoxide of iron .. ..   0·950
        Manganese   .. ..   1·860
        Potash   .. ..   0·089
        Soda   .. ..   0·138
        Sulphur   .. ..   0·030

Slags from furnaces at Lemington, Northumberland, making Bessemer pig-iron, by Proctor.

	1.	2.
Silica	31·83	32·21
Protoxide of iron	0·25	trace
Lime	32·39	48·62
Alumina	9·89	12·64
Magnesia	3·89	3·23
Protoxide of manganese	1·23	..
Sulphur	1·44	..
Sulphide of calcium	..	3·19

Buffalo, New York. Anthracite furnace using red hematite ore.

Protoxide of iron	0·55
Silica	39·35
Alumina	13·86
Lime	37·63
Sulphate of magnesia	3·65
" manganese	2·40
Sulphur	2·42
Phosphorus	trace

At Middlesborough and other parts of England the conversion of slag into bricks and blocks for building has been conducted with success. For foundations on which heavy engines and machines are fixed it has been for a long time used. For such purposes a varying proportion of caustic lime has been mixed, as well as gravel and broken fire-bricks, with satisfactory results. A specimen of a piece of concrete so prepared for a foundation to receive the weight and shock of a large steam hammer is in the possession of the author, and it compares favourably with the best prepared Portland cement concretes. There is, however, some discrimination necessary in selecting such slags as exhibit on analysis the smallest percentage of sulphur, or its sulphates, or sulphides. For foundations in dry situations such precaution is not so necessary, but when building houses or similar structures an excess of these deleterious substances would result in ultimate disintegration.

At Osnabrück an artificial stone is prepared and moulded into blocks from the George-Marie iron works.

The slag in this case is dealt with direct from the blast furnace, and is brought into contact with cold water, by which operation it is granulated. In this state it receives an addition of from 15 per cent. to 25 per cent. of lime, and is then passed through an ordinary mortar mill. This mixture, from its want of cohesiveness, cannot be treated in the usual brick-forming machines, but is submitted to a treatment somewhat analogous through the agency of a specially constructed machine of the following description.

A horizontal box, from either end of which a piston is pressed forward to its middle or centre, where is situated the hopper by which the machine is fed with the prepared material. When the exact quantity required to make one of the blocks has passed into the box, both pistons are pressed forward, thus completing the final process of compression. This being accomplished, one of the pistons is drawn back, the other following with the moulded brick until it reaches an opening, into which it drops, and from thence it is carried away by bands or other similar arrangements to the stacking grounds. The box and its accurately fitting pistons are of the required form or section it is determined to make the brick or block, and may be altered according to circumstances.

Bricks thus moulded from such a preparation do not require a larger amount of pressure than may be necessary to expel the air—at least theoretically; for practically no brick-forming machine has yet accomplished this desideratum, as the indurating process is brought about by the chemical formation of silicate and carbonate of lime.

At these works there are five presses in operation, whose combined produce amounts to 30,000 blocks per day. One hundred tons of slag is thus consumed in the fabrication of a day's production. During the year 1873 six millions of bricks were thus made and used in the construction of public and private buildings. The cost was 27s. per 1000, and the weight about $7\frac{1}{2}$ lbs. each. It appears to the author that the cost is high, and could be done on more favourable terms in this country.

On a careful examination of the various analyses of slags it will be seen that in many of the specimens it would not require much addition of lime, and certainly not any very expensive treatment to convert them into a good Portland cement. At all events, it is desirable that, when practicable, such a conversion, or for that of buildings, blocks should be entertained. It is obvious that any operation or treatment of the slag, direct from the furnaces, would result in great advantage to the ironmaster, whose interest is certainly mixed up with its utilization.

With such a wide-spread and almost unlimited field from which slag can be obtained, the most doubting need not entertain any fear of its being exhausted. With the exception of Ireland, inexhaustible deposits abound from the north in Scotland to South Wales, and between these extreme points a wide field for selection exists. The comparatively recent discovery and employment of the argillaceous iron ores of Northamptonshire has led to the erection of blast furnaces within seventy miles of London. Ireland is happily circumstanced in having abundant deposits of iron ores, but the absence of cheap fuel for its profitable conversion necessitates its shipment to Scotland and England. Its export of ores now reach nearly 150,000 tons per annum.

WASTE FROM LIME BURNING.

Whether we regard the waste accumulated in the chalk districts of the south principally in its natural condition of carbonates of lime, or whether we examine the deposits of lime and its hydrates from the lime works in the carboniferous and other formations, their extent and diffuseness is remarkable.

In the Medway valley Portland cement making has reduced this waste to a minimum, but with this exception, these deposits are continually increasing, unless we may notice the slight impression made in the carboniferous limestone district of North Derbyshire, where there is being made, under a patent granted to the author, a good marketable Portland cement, which finds a ready sale in Lancashire and elsewhere.

The analysis of this waste is, as might be imagined, somewhat variable in character, owing to the fluctuating quantity of shale and unconsumed coal with which it is mixed. An average sample, analyzed by W. F. Reid, gives the following results:

Sand	9·30
Sulphur	1·11
Silica	4·00
Carbonic acid	3·33
Lime	52·43
Magnesia	0·55
Alumina and oxide of iron	3·19
Water	26·24

This material is essentially a hydrate of lime, and from its absorbing so much moisture requires, before treating it as a cement ingredient, the necessary degree of heat to render it sufficiently desiccated to be operated upon by the reducing machinery. The carbonic acid in this analysis is due to the absorption from the atmosphere after having laid to *waste*, and its amount would be fluctuating according to the time of its exposure. In the neighbourhood of the deposit from which this specimen was obtained there may be procured examples in a state resembling quicklime to those as hard as rock, which can only be separated from the mass by the aid of a hammer. Indeed, where an improvement was made in a turnpike road passing through an ancient waste, the operation of quarrying had to be resorted to.

Although this waste has been for the first time thus utilized, its use as an ingredient in mortar making has been for a long time recognized, and in Manchester and other parts of Lancashire and Yorkshire large quantities are employed for this purpose. When passed through a carefully controlled mortar mill, in combination with good sand, it makes an excellent mortar. To obtain the fullest benefits, however, it is necessary that the burnt shales (product of the inferior quality of coals used in the process of lime making) should be as finely comminuted as possible. The new waste—which is practically caustic lime, combined with the shale and unburnt stone—is generally used, although for some purposes a proportion of the old partially recarbonated waste is mixed. In the numerous deposits within the Buxton district wastes exists ranging in quality from that newly deposited from the kilns in a free or caustic state, to those probably exceeding a century in age, and almost as hard as the stone from which they were originally obtained.

Barn and cottage floors are commonly made of this waste, and when carefully rammed and made perfectly compact in texture they are durable, and, especially in the mountain districts, are in much favour. Floors of this or a similar kind are common in the south of England. The earthen barn floors on the Cotswold Hills are thought to be superior to any, except those of oak plank. The materials used are the calcareous earth of the subsoil, mixed with freestone chippings, in equal quantities, and worked up dry; the surface being levelled, it is beaten with a flat wooden beater, similar to that used by gardeners for flattening lawns. These floors never crack, and are soon ready for use.

Virgil, in the 'Georgics,' while describing the various implements of the Roman cultivator, and directing his attention to the best means of rendering his occupation successful, amongst other things mentioning the necessity of a good granary floor, says:

> "First on the floor be all thy art bestowed,
> Level its plain beneath the roller's load;
> The clay well wrought should feel thy griping hand,
> And binding chalk consolidate the land,
> Lest chopped the earth, by arid dust subdued,
> Should nourish weeds, and various vermin breed."

It is somewhat unaccountable that so much waste has thus been permitted to accumulate, and its continuance after better systems of burning had been elsewhere adopted with success. There is now, however, a desire to adopt a more rational system of lime making, and when Hoffman's or other kilns constructed on a sound scientific basis have been substituted for the antiquated running kilns the quantity of waste will be but trifling. Indeed, theoretically, with a well designed and rationally managed ring kiln there should be no waste at all. There is probably in some districts much difficulty in substituting any of the improved forms of kilns for the old ones. One strong reason is the necessity for burning a local and cheap coal, which could not be used in those of the ring-kiln type as well as where a limited quantity only is needed, as a chamber kiln, from its requiring to be kept continuously at work, would provide too much lime for local wants. In a district well supplied with railway accommodation, this might be in a degree overcome by concentrating the various small works into one convenient centre, whereby the latest and most approved appliances could be brought to bear with advantage.

These waste heaps are generally in close proximity to the other ingredients, in the form of shales and clays. At present their cost is merely nominal, and even supposing their further increase to cease, there is an abundance in many districts to supply the cement maker during this century at least. That good Portland cement is being made from this source establishes its value in that direction; but, like every other supply of the carbonate or hydrate of lime, it must be converted in the locality in which it lies, as no expensive carriage for these materials can be tolerated.

Alkali Waste.

This waste exists in considerable abundance, but its locality is limited to the seats of the alkali manufactories, the more important being those in Lancashire, Northumberland, and Scotland (Glasgow). The lime waste last referred to has no injurious influence in its character, and the damage incurred by its deposits is confined to the destruction of vegetation on the soil which it overlies. In that produced from the alkali waste a very different result is brought about, for not only is the atmosphere rendered noxious in and about the locality of manufacture, but vegetation for long distances is seriously checked. This damage has been at all times a source of great loss to the manufacturer in meeting the compensation demanded for the injury which his operations have occasioned, and has in some instances led to the outlay of large sums in the erection of high chimneys, which have failed to remedy the evil, for while partially suppressing the nuisance in the immediate vicinity, they have created a more distant nuisance, which until then had not been felt. The legislative measures, now become very stringent, will, it is hoped, cure this evil; and in doing so, prove a boon to the manufacturers themselves by the economization of the waste gases, which, while being absolute loss from a manufacturing cost point of view, contribute by their wide-spread distribution external loss of a very expensive character.

Some attempts have been made to render this waste available as an ingredient in the manufacture of Portland cement, but owing to the large amount of sulphur present in such heaps no appreciable success has been attained. The author was consulted some years ago by one of the largest alkali manufacturing firms in the St. Helen's (Lancashire) district, with the view of utilizing their waste, and made a series of experiments to ascertain its suitability for Portland cement making. It was found, however, that the sulphate of lime prevented the fabrication of a perfectly hydraulic cement, although for ordinary purposes, such as concrete and mortar, it might have been suitable. This firm at that time were making large quantities of waste, and difficulty had arisen in the easy and cheap obtainment of suitable and convenient depositing ground, thereby involving a large outlay in carting the waste from their works. Had these experiments been regarded as at all satisfactory, the firm were prepared, and it would have paid them, to erect Portland cement works.

Since the time of these experiments de-sulphuring processes have been resorted to for the recovery of the sulphur, and it is possible that better success might now attend the treatment of the purified waste. It is not claimed for any of these processes that the whole of the sulphur is recovered, but the amount extracted leaves a much better chance of making a good cement.

In the localities in which these wastes occur frequent use is made of them for various purposes, such as floors and foundations, for which in ordinary cases they are well adapted. But inasmuch as the sulphur prevents successful treatment in a cement-making direction, we will not recommend any extensive experiments at present.

One very favourable circumstance in connection with this waste is that it invariably occurs in districts well situated for the supply of fuel and with unusual facilities for shipping. Of course in selecting sites the alkali manufacturers gave special attention to the cheap supply of their raw materials and an easy and cheap outlet for the distant markets it was meant to supply. Owing to the necessity for using as pure a carbonate of lime as possible, some of these manufacturers labour under great disadvantage. The Derbyshire limestone is taken to Gateshead-on-Tyne at a great transit cost. Such a fact speaks highly in favour of the purity of this limestone, for in Northumberland and adjoining counties limestone of fair quality exists in abundance. Of the many attempts made to utilize the waste sulphur, none may be regarded as perfectly successful, but as they approach that point, the more suitable or acceptable will the residue become for the cement operator.

Alum Waste.

The manufacture of alum from the various aluminous shales in the north of England and Scotland is a most important industry, and like the alkali trade results in the accumulation of enormous waste heaps. An old and important seat of this industry exists in the neighbourhood of Glasgow, at Campsie and Hurlet, where great mounds of exhausted shale attract the attention of the visitor. The following are analyses of this waste, No. 1, from Hurlet, by Professor Penny, and No. 2, from Campsie, by Dr. Muspratt:

	1.	2.
Silica	44·0	38·40
Alumina	22·2	12·70
Sesquioxide of iron		20·80
Oxide of iron	11·4	
" manganese		trace
Lime	2·2	2·07
Magnesia	0·3	2·00
Sulphuric acid	5·7	10·76
Potassa		1·00
Water	14·2	12·27

In the neighbourhood of these deposits, but especially at Campsie, are to be found limestones of good quality which could readily be combined in the production of a good Portland cement.

ARTIFICIAL SOURCES OF CEMENT-MAKING MATERIALS. 133

Soap Lime Waste.

The author upwards of twelve years ago converted a considerable quantity of this material into Portland cement in combination with Medway clay, and found it a profitable conversion, but, owing to the supply being limited, he does not recommend its use for cement making except when the purpose for which it may be required is a purely local one. The process of conversion is easy, owing to the favourable condition (in point of texture) of this material.

Gas Lime Waste.

The lime used in this process is for the purification of the gas, and the quality resulting depends in chemical value pretty much on the condition and quality of the coal from which the gas is obtained. Sulphur, the declared enemy of cement makers, abounds in the impure coals, and the following analyses show the fluctuating value obtained:

No. 1.—By Dr. Graham.
" 2.—By Dr. Voelcker (at 100° C.).

	1.	2.
Hyposulphate of lime	12·30	15·19
Sulphate of lime	2·80	4·64
Sulphite „	14·57	..
Carbonate „	14·48	49·40
Caustic lime	17·72	18·23
Sulphur	5·14	..
Silica	0·71	0·28
Water	32·28	7·24
Oxide of iron, clay, &c.	..	2·49
Magnesia and ammonia	..	2·50

Analysis of lime waste from wood-gas making, by Dr. Reissig:

Carbonate of lime	69·16
„ magnesia	5·60
Caustic lime	9·90
Oxide of iron and alumina	3·82
Silica	3·56
Volatile organic substances	1·75
Water	5·99

With this more curious than useful analysis we will close this chapter, and trust that the information therein intended to be conveyed may result in practical good and direct attention to every substance from which cements are attainable, whether locked up by nature or hidden from our notice by man's carelessness and waste.

CHAPTER IV.

CONSIDERATIONS WHICH SHOULD INFLUENCE THE MANUFACTURER IN SELECTING A SITE FOR THE CEMENT WORKS.

In the preceding chapters, treating of the different kinds of raw materials which are or may be converted into Portland cement, it has been shown how their various characteristics and peculiarities affect the question which is being discussed. It will seldom be found that any two of the distinct geological formations exist together in sufficient quantity to interfere with the desired choice, and it is therefore useless to guard the intending manufacturer against the danger of selecting any point which is not likely to secure a supply of raw materials for a considerable number of years, unless the works are only to be of a temporary character for the fulfilment of a specific object or purpose.

Where cement works are to be erected for the special supply of cement during the construction of some important engineering or architectural works, such conditions cannot prevail, as it is not likely that the solution of the site of works of any importance would be influenced by their nearness to or remoteness from the vicinity of the raw materials. The subject of cement making, however, has become a consideration with engineers, and some of the most eminent in the profession have already declared for its necessity by the construction of cement works where large drainage, water, and harbour works are being built.

The first consideration to be regarded, for whatever purpose the works may be required, should be of course the control of the raw materials, so as to command their delivery at the works at the lowest cost. When possible the manufactory should be built on ground where one or other of the raw materials exists; preferably on the carbonate of lime, as it forms the largest ingredient in the raw material mixture. While, however, having due regard to this essential, it must not be overlooked that the command of the other ingredients is also of equal importance; on a site commanding carbonate of lime only and remote from the clays or shales, involving thereby cost of carriage, works could not be carried on profitably unless such a disadvantage is counterbalanced by some other important consideration, such as nearness to cheap fuel, or facilities for the transport of the manufactured article.

It is desirable also to consider how the works are to be connected with the best and cheapest railway, or other means of communication with the neighbourhood or market which it is intended to supply. Railways may be regarded as the most convenient. Canals are best adapted for the cheap conveyance of cement and the materials from which it is made; but unfortunately the majority of these useful arteries of commercial communication are in the hands of, or controlled by, railway companies, and thus practically beyond the reach of cement manufacturers. River and sea communication may therefore be regarded as the most acceptable mode of securing a cheap and accessible means of transit.

It happens, however, also that in this country the most valuable deposits of carbonate of lime occur in unlimited abundance on the sea shores and on the river banks. The question of labour is one which should have due consideration, and should therefore weigh in the scale with those of more importance. As the materials of each locality will more or less influence the system of manufacture which will have to be adopted, it should not be overlooked that the process of one district cannot influence or be acceptable in another.

For instance, a practical cement maker, unacquainted with the technical character of the business, trained in the chalk district, would be comparatively useless in the lias or other distinct geological division of the carbonate of lime series. The very initiatory processes differ in character, and so does, in fact, every stage of its manipulation, from the first reduction of the stone and shales to its final destination in the kilns. There is only one branch of the business, however, which is common to all districts, and that is the process of grinding. This is a purely mechanical operation, and its accuracy of result is in nowise dependent on the locality in which it is performed. The science or art of stone dressing is of ancient origin, and only requires an exact knowledge, acquired by intelligent practice, of the different dresses (diagrams of the millstones) which shall secure the most economical results.

The writer is not so sanguine as to think that any large works equal in extent to the leviathan manufactories on the Thames, at Boulogne, or Stettin are likely to be erected in the interior of our island; but from his knowledge of the growing wants of concrete structures, he is convinced that many works of ordinary extent would find profitable results in supplying the cement which such a demand engenders.

Although perhaps, as a rule of pretty nearly general application, the advantages are in favour of a manufacturing site on or immediately commanding a supply of carbonate of lime, it should not be overlooked that even such an advantage may be obtained at the sacrifice of other considerations.

In weighing the various advantages of a favourably circumstanced carbonate of lime locality, it would be as well to enter into a careful calculation in some such way as the following. As carbonate of lime forms generally a proportion of three to one by weight to the clay, we should examine the cost of this material to a manufactory from which is to be issued a given quantity of cement per week. And presuming the supply of clay or shale obtainable only at a distance, the calculation is not a difficult one to ascertain in which position the site of the works is to be selected. This point of relative cost being settled, the next is that with reference to the cost of fuel, which should be considered having regard to the weight required for the conversion of the required quantity of raw materials into the manufactured article. The weight of the fuel for all purposes varies, according to the process adopted and the materials dealt with, from 10 to 20 per cent. of the gross natural weight of the raw materials operated on. This weight of fuel, it should be recollected, entirely disappears during the manufacture, and forms therefore no new element in the consideration of weight to be further dealt with. We have now only the manufactured cement, which should in its resulting condition be about 60 per cent. of the gross weight of the raw materials, of course without the fuel, when deprived of their latent and acquired moisture. The remaining consideration is one of comparative insignificance, as it only affects the cost of carriage of the manufactured article. Due regard, however, should be had to the accessibility of the site selected to the markets which it is intended to supply. It is needless, however, to say that a cement works in an isolated situation, at a distance from railway, canal, river, or sea communication, would be misplaced and not likely to prove remunerative.

The necessary care has not always been displayed in the selection of the sites for many existing cement manufactories, and the author is familiar with several that struggle against the evils entailed by a primary error of this kind. Intending manufacturers should never lose sight of the necessity for commanding at a nominal cost at least one of the raw materials, and, by preference, of the carbonate of lime, for the reasons before named. Works so circumstanced as to require to be supplied from a distance with all the raw materials, labour under disadvantages which the most careful management cannot hope to overcome. During periods of prosperity and high prices, the full effect of an ill-placed or badly selected site are unfelt. It is only when those frequently recurring periods of excessive competition arise, and when the make of cement exceeds the demand, that the utmost care is often unable to avert the losses which an over costly supply of raw materials is sure to bring about. Too much attention cannot therefore be given in the initial calculations for the selection of a site, for it is difficult afterwards to recover from an error of this kind.

In confirmation of these observations, it is only necessary to examine the localities of the most successful and largest works in England. They are all constructed on the rivers or coast, and invariably according to the degree with which they command the raw materials have their success been assured. It should be observed here in this section of our argument, that it would be difficult to equal in elegibility the site which the Portland cement works on the rivers Thames and Medway command, not only as regards the supply of the raw materials, but a cheap transit for the cement when manufactured. Indeed, the happy combination of advantages which these favoured works command, has undoubtedly tended in a high degree to the advancement of the trade and influence of Portland cement in all countries. Although Wakefield must be regarded as its birthplace, the fostering nursery of its early struggles and prosperity is due to the influence of the London manufacturers.

CHAPTER V.

OBSERVATIONS ON THE SELECTION OF THE RAW MATERIALS.

In the preceding chapters we have endeavoured—in what may perhaps be considered a too lengthy manner—to point out, in a geological and mineral sense, the various simple and compound materials capable of utilization by the Portland cement manufacturer. We will now proceed to discuss generally the usual methods of estimating their value and ascertaining their suitableness for cement making.

Unless the manufacturer has fitted himself for a task of this nature by a previously well-considered study of chemistry, in so far as it can fit him for the duty of appreciating the merits and valuing the properties of the materials with which he has to deal, it would be more advisable for him to seek the assistance of a professed chemist to guide him at starting. The analysis need not necessarily be of an exhaustive character, but such as may safely act as a basis on which to start fairly on the initiatory experiments. It is not absolutely necessary that the value of other than the three leading and indispensable ingredients, carbonate of lime, silica, and alumina, should be ascertained, unless the presence of sulphur in any of its varied forms should be present in the sample submitted to analysis. Much trouble and loss of time would be averted by a preliminary examination of the physiographic character or aspect of the stones, shales, or clays. A careful observation of their character of combination would probably be the means of saving considerable cost in their ultimate treatment. Of course much will depend on the process of manufacture selected, and the materials to be operated upon will greatly influence the operator in that selection. While some materials admit of being converted into cement, by either the wet or dry process, many are only capable of profitable conversion by the one process or the other, and some again admit of such treatment as may be considered a combination of both methods. For the better elucidation of this part of the question we will call attention to the materials themselves. First, the chalk, as being the most important, simplest, and best-known material. Chalk, unless where metamorphosed by volcanic action, as at the Giants' Causeway and elsewhere, is an amorphous mineral and capable of being rendered sufficiently solvent in water without much trouble. The impurities with which it is more or less combined are generally of such a harmless character as to render their existence in the preparatory reduced mass a matter of comparative indifference. There is no necessity therefore for any very accurate chemical examination of this material. Perhaps the simplest and best approximate test is to examine its physical characteristics by reducing a small quantity in water and decanting it until the residuum can be estimated. The various qualities of chalk will display very different results, but it may safely be assumed that the insoluble residuum is not of an objectionable nature, or calculated to endanger the quality of the cement, if all the other necessary steps in its manipulation are accurately performed. In further illustration of the harmlessnes of any sediment from the chalk, it may be shown that if it be reduced by the wet process its passage into the backs is rendered impossible when the proper precautions hereafter to be explained have been complied with. And again, should it be used by the dry system, its reduction by the millstone or grinding agency would also effectually prevent its exerting any injurious influence on the final result. In short, chalk is not to be regarded with hesitating suspicion by the cement maker, for it is the simplest and kindest of the materials used in cement making. Where a choice exists it is better to select the smoothest in texture and the quality most easily reduced to a state of comparative solubility. In conjunction with the chalk from the several formations, river-mud or clay is invariably used, and, as before observed, its careful selection is necessary. A sedimentary deposit so peculiarly formed and subject to so many fluctuating conditions results in a variety of qualities of the deposited material. The finer and more suitable qualities are readily soluble, and unless they are comparatively free from any sediment of fine silt or free sand, their use is attended with some danger. In being worked with the chalk it is closely incorporated with the combined semi-liquid, and, being generally fine and of light specific gravity, it escapes through the sieves to the back, where its presence is most objectionable. It is better therefore to exercise a little extra care in selecting the raw material, for however perfect in theory the subsequent arrangements may be for intercepting the sand on its passage to the backs, they seldom in practice attain the desired object. Any clay for use under the wet system should not have a larger amount of sand in its composition when dug than 5 per cent.

The next in importance and value is the limestones and shales of the lias deposits. Their chemical value is, however, so fluctuating that accurate analyses of each distinct band or layer is indispensable. As this class of materials are only capable of conversion into Portland cement by the dry process, any objectionable ingredient must remain in the mixture if not challenged before selection and admixture. In this deposit a large amount of iron pyrites is usually found, and when permitted to remain in the mixture dangerous properties become developed, and the hydraulicity of the cement is rendered uncertain. The limestones of this formation are almost entirely free from this objectionable ingredient, and therefore need no care in guarding against this species of danger. The shales display on fracture, when freshly quarried and after their exposure to the air, on disintegration, the shining crystals of the iron pyrites. The various beds indicate a very fluctuating quantity of iron pyrites, and in some of the beds it is absent altogether, or at least exists in very inappreciable quantities. As the shales in the lias deposits preponderate and require removal in order to obtain the limestone, it is easy to select the best qualities of shales, which are generally found to be those smoothest in texture. The natural admixture in these shales is so perfect that little if any risk arises from any inconvertible ingredient. It is usually found that much difference in analytic value exists in the separate beds, showing that during the period of their formation the sediment deposited was variable in character. Each separate leaf or bed has special and distinct physical and chemical qualities.

A compound treatment of these shales, after becoming disintegrated by exposure to the atmosphere, might be beneficially employed. They could be put through a wash-mill, and by this means purged of the iron pyrites—for they generally exist in mechanical combination—and the expense of this operation might frequently be repaid from the value of the pyrites recovered, for in some shales it exists in considerable abundance. The higher specific gravity of the pyrites would readily, in a semi-liquid mixture, gravitate to the bottom of the mill, and thus facilitate the passage of the fluid shale to the mixing machine for combination with the carbonate of lime.

The oolitic limestones and all the other varieties of carbonates of lime differ in their mineralogical aspect or condition, and according to the extent of their crystallization will the treatment which they require vary. In the handling of amorphous materials it is only a question of simple mechanical operation to reduce the varying indurated chalks, clays, and shales, so as to fit them to perform their required task of free and accurate admixture. In dealing with hard crystallized rocks the task is much more onerous, for, however finely you may reduce the material, you cannot annihilate the crystal; no matter how small the atom, it is still a crystal, and your utmost sources of mechanical reduction fail to effect its complete extinguishment. The mechanical expense of this operation is therefore considerable, and, however accurately performed, it may never succeed in the realization of a powder fine enough or sufficiently soft to ensure the perfect chemical combination of the various materials. To accomplish what must be regarded as the most important part of cement-making operations, it is desirable, to ensure the success of this initiatory stage of manufacture, to first convert the limestone into lime by the ordinary process of burning. That alternative thoroughly overcomes the objection to the crystalline character of the mechanically reduced stone, and favours the subsequent process of amalgamation, which will be rendered more intelligible in the chapter treating of the mixing process.

Whatever system of treatment is resorted to in the preparation of the raw materials, the primary object of their amalgamation is to ensure their favourable capacity for the all-important chemical operation performed in and through the agency of the kiln. There is doubtless—and its importance also cannot be too strenuously insisted upon—much accuracy required in ascertaining the exact proportions of the different raw materials, but that process will have been in vain if the ingredients are not sufficiently fined down to enable the delicate and final operation of the kiln to be performed. The simple unconditional mixture is not alone all that is required, but the necessary vigilance must be insisted upon so that its quality should be that of a fine and almost impalpable powder, otherwise the final cement result is unattainable.

You will not often find all the raw materials in one locality, but it is most advisable that the required technical skill should be forthcoming in determining the best available materials to be selected for cement-making operations. Many of the failures in making good Portland cement are due to the clumsy and ignorant treatment of the materials, and not to their natural unfitness for the purpose. Although the question of cost of the raw materials must always form an important element in the consideration of the expense of manufacture, still it must not be pushed too far, for in reality the original cost of the raw materials forms but an inconsiderable item of the whole cost of manufacture.

In the first efforts at Portland cement making by Aspdin, he was confined in his selection of materials to the immediate locality of his manufactory. The subsequent adoption of the London district as the manufacturing point resulted in a successful treatment of the chalk and river clays. The operations, however, were of an inaccurate and hap-hazard character, and until recently were entrusted to the ordinary workmen who could produce the most perfect mixture of the two simple materials, chalk and clay. The resulting process was again subject to the control of the man who could best apply the required amount of fuel for its ultimate calcination. It was found, however, that some more intelligent supervision was required to ensure accurate results, and to produce such a mixture as could be burnt by an unvarying and fixed amount of fuel. Hence in recent years the institution of tests on the fluid mixture flowing from the wash-mills, thereby ensuring a more even system of manufacture. Still it can of necessity be but approximate, and far short of what may be regarded as a technical or scientific test of the operation. This subject will be more fully dealt with in the following chapter.

In converting chalk and clay into cement, the standard required in the operation—under the present system at least—need not be regarded as a high one, and the slovenliness of the process, so long as it results in profit, is not regarded as objectionable. It involves, however, a great waste of labour, if not of material, the truth of which we shall endeavour to prove before the close of our argument.

In dealing with the mountain or carboniferous limestone, there will not be found much variety in the physical character of the stone. It may be regarded as a sub-crystalline rock and easily fractured, which property becomes an advantage in the final treatment when being powdered in the horizontal or mill stones. It is preferable to select a silica amalgam of as hard a nature as possible, for the nearer the two ingredients approach each other in texture, the better and more economical will be the results obtained in grinding. In some districts of this formation, especially when it has been influenced by the "toadstone" and "basalt dyke" intrusions—a species of "chert" is found of a highly quartzose character. A similar metamorphism occurs in the toadstone's contact with the gritstone formation, although not to so great an extent.

Where no higher knowledge can be commanded by the experimenter than that we have described fully in Chapter II. we would recommend a careful perusal thereof before determining on the final selection of the raw materials. This observation has special reference to all the materials excepting those of chalk and river clays, which are so well known and understood.

In the igneous rocks much difference of value may be found, but the aqueous or sedimentary deposits are, generally speaking, reliable. In some cases it may be found that the same beds of limestones not only differ in chemical value, but also exhibit considerable variety in their physical aspect or condition. Their thorough reduction and consequent mixture overcome much of the risk attending such characteristics, and attention is only directed to it for the purpose of cautioning against the chance of being led into error at the time of examination. There should be a series of tests when such differences are apparent.

CHAPTER VI.

THE ESTIMATION OF THE RAW MATERIALS.

HAVING thus fully entered into the geology and mineralogy of the raw materials, their sources of origin, and selection, we shall proceed to discuss the best and most approved mode of ascertaining their mechanical and chemical value.

We shall first describe the apparatus used in estimating the physical qualities of clays by mechanical estimation.

The physical properties of clays are almost as important as the chemical, and in many cases equally variable. According to the interesting researches of Dr. Aron, of Berlin, one of the best and most reliable authorities on this subject, pure clay is composed of globular or spherical particles. These particles are smaller than the mineral detritus with which they are mixed, and can be separated from the latter by a very slow current of water. The porosity and plasticity of several different clays in their natural state may differ widely, but will be almost identical in the case of the products obtained from them by careful washing. The space between the particles is in the case of each washed clay almost the same, proving that these particles must be of some regular shape, and the microscope shows this shape to be spherical. The plasticity of clays seems to depend upon the proportion of such globular particles contained in them.

For practical purposes, however, such accuracy in the mechanical analysis of clays is seldom necessary. It generally suffices to determine the amount of sand present and the size of the grains.

This can best be done by Schöne's apparatus, Fig. 1,

which admits of a degree of precision unattainable by any other instrument.

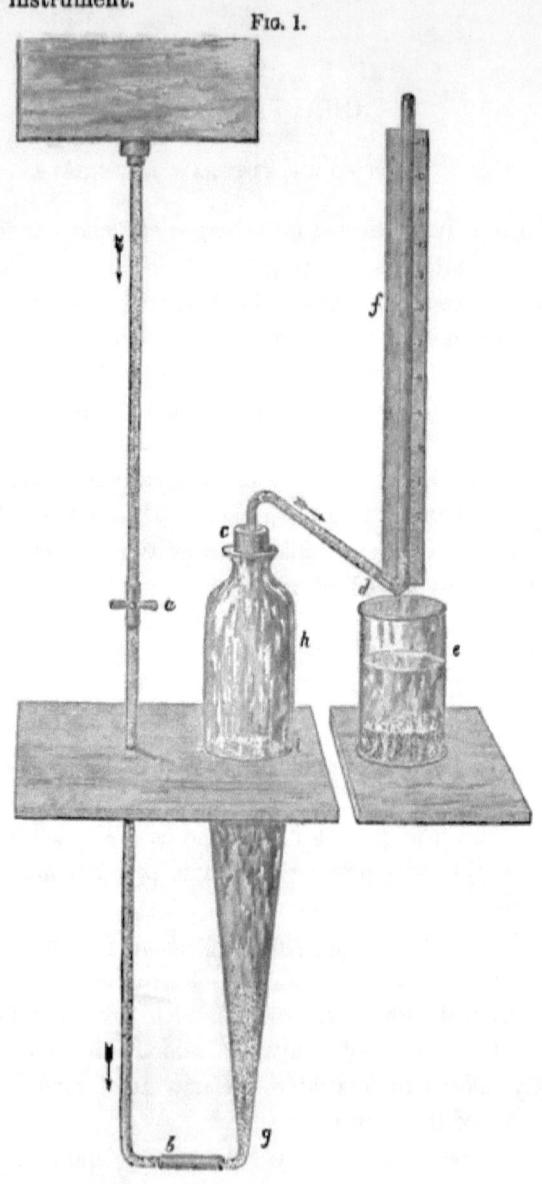

Fig. 1.

The most important part of this apparatus is the glass vessel, part of which, h, i, should be perfectly cylindrical. This vessel communicates with a cistern, the supply of water from which can be regulated by means of a. The velocity of the current can be measured by means of the scale f, up which the water will rise in proportion to the quantity passing through. The clay to be washed is placed in the apparatus at c; the cork is replaced, and the water turned on. The water containing the fine particles passes through the opening in the lower bend of the tube at d, and may be collected in a beaker glass or other suitable vessel. The coarseness of the particles will increase in proportion to the height of the water in f. Should the clay contain any coarse sand, a small disc of wire gauze inserted at g will prevent the tube from being choked. Most clays must be prepared before placing them in the apparatus. Those containing carbonate of lime may be boiled in a weak solution of hydrochloric acid; in any case it is advisable to boil in water till a perfectly homogeneous mass is obtained. Any very coarse particles should be removed by means of a small sieve of wire gauze.

In general, the less sand a clay contains, the better is it adapted for cement making. Very fine impalpable sand may, however, be present in considerable quantities without exerting an injurious influence on the cement, especially, if the percentage of alkalies contained in the clays is considerable. Cement made through the agency of arenaceous clays requires a more extended period of burning as well as the exercise of greater care. Clay containing more than five per cent. of coarse sand, and which to the touch feels *gritty*, should undergo a preliminary washing, so as to purge it from so objectionable an ingredient.

During the early period of Portland cement making, none but the softest and most plastic clays were used, but since the introduction and development of the modern dry process, shales, slates, and basalts have been used with very considerable success. Next in injurious influence to sand is the presence of sulphur in the clays, or other silica and alumina producing ingredients. It generally occurs in the form of pyrites or gypsum (sulphate of lime). Gypsum can be readily detected by the shining or lustrous character of its crystals, especially on the cleavage planes: pyrites is sometimes so finely disseminated throughout the mass, as to necessitate its estimation by accurate chemical analysis. If the percentage is considerable a small quantity roasted over the flame of a spirit lamp will give off a strong smell of burning sulphur, and where it exists in a proportion equal to 2 per cent. of sulphuric acid, it renders a clay in its natural state unfit for cement-making purposes. By a preliminary treatment, however, its injurious influence may be avoided.

Carbonate of lime is very frequently found in clays, more especially in the gault and lias, and other more recent formations. Its presence in clays, and when finely distributed, is not of much consequence; but if it occurs in irregular mechanical distribution, great care is necessary to protect the mixture from an improperly balanced proportion.

Oxide of iron is found in almost all clays, and when it occurs in a high degree, it tends to increase or deepen the colour of the cement; it is never, however, present in such quantities as to materially interfere with the strength of the cement itself.

Alumina, when in excess in a clay, impairs the indurating value of the cement in the making of which it is used. Aluminate of lime possesses excellent hydraulic properties, but the temperature necessary for its formation is much higher than that at which silicate of lime is produced. If therefore a clay contains an excess of alumina, part of the silicate of lime will be overburnt before the whole of the alumina can enter into combination with the lime.

Organic matter is seldom present in clays in any large proportion; but it may be of importance to some manufacturers who operate in populous neighbourhoods, as it is the cause of those offensive odours emitted by cement and brick kilns. The smell caused by the burning of coke by itself will be found to differ greatly from that of a cement kiln, and if clay is heated in a retort to a high temperature, it gives off vapours with an odour *sui generis*. Chalk also contains minute quantities of organic matter, and some limestones, more especially those obtained from the carboniferous and lias formations, are highly bituminous.

Originally when the quantity of clay required for cement making was very limited, the obtainment of the purest varieties was a matter of little difficulty. The increasing demand, however, without an increase of the sources from which supplies could be obtained, produced a more vigilant attention to the varieties and peculiarities of the esturian deposits. The *mouth test* of the old-fashioned cement washer could no longer secure what was required, and the estimate of the gritty effect produced against his teeth was abandoned as the then recognized challenge of its purity.

In the generality of cases, a perfect chemical analysis of a clay will determine its suitability for cement making, but the above mechanical estimation is offered as a ready means of detecting unsuitable clays by the aid of this apparatus. Much loss has been sustained by practical experiments in the manufacture of cement, which never would have been undertaken, for if a qualitative analysis had preceded them, the unfitness of the materials would have been ascertained. Accuracy of analysis need not necessarily precede experiments, but it should be performed before embarking in much outlay, for not only will it prove the suitability of the raw materials, but indicate the proportions of their admixture.

In the search for raw materials, it will save much time to have at hand a ready and speedy means of estimating the carbonate of lime value of the materials under investigation. The apparatus we are about to describe possesses in a high degree this advantage, and may become in the hands of the cement maker a useful instrument, not only for proving the raw materials, but as a guide in testing the proportion of their mixtures.

This apparatus was first designed by Dr. Scheibler, of Berlin, to estimate the amount of lime contained in the animal charcoal used in the process of refining sugar, and may be called an apparatus for the volumetric determination of carbonic acid.

When all the lime in any substance exists in the state of carbonate (no other carbonate being present), its quantity may be calculated from the amount of carbonic acid evolved on the addition of some stronger mineral acid, such as hydrochloric acid. Carbonate of lime is a substance of known definite composition, containing 56 per cent. lime and 44 per cent. carbonic acid. If therefore we find in a limestone 36 per cent. of carbonic acid, then the percentage of lime will be $\frac{36 + 56}{44} = 45 \cdot 8$ per cent., and the percentage of carbonate of lime $36 + 45 \cdot 8 = 81 \cdot 8$ per cent.

The substance containing carbonate of lime should be pulverized and dried on a water-bath. When cool, weigh off $0 \cdot 5$ gramme of the dry powder. If a chemical balance and weights are not accessible, a weight should be procured weighing exactly $0 \cdot 5$ gramme, and the substance can be as easily weighed in a pair of ordinary apothecary's scales with sufficient accuracy for such purposes as those under discussion.

154 SCIENCE AND ART OF PORTLAND CEMENT.

Fig. 2 represents this apparatus, and the mode recommended for ascertaining the carbonate lime value of any substance of which it forms a part, is as follows:

FIG. 2.

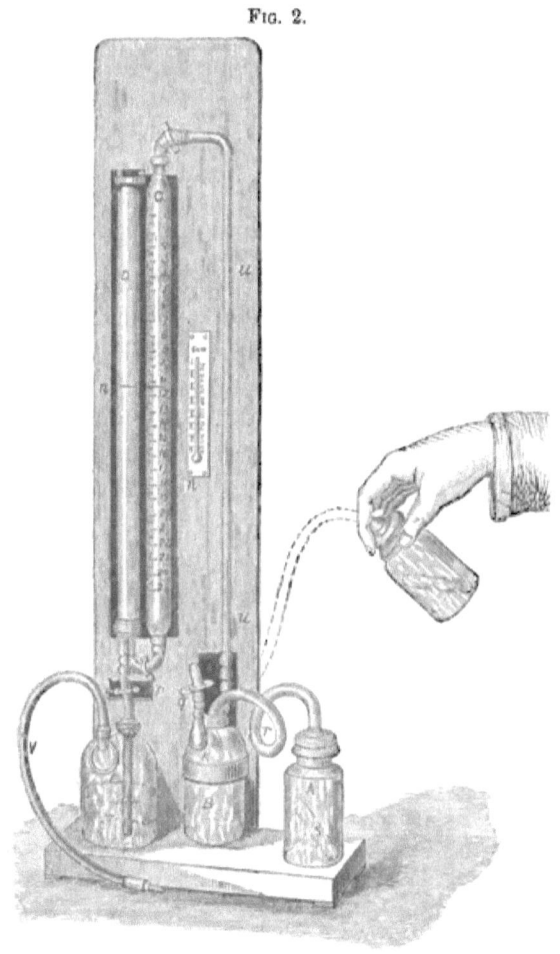

Place the weighed substance carefully in the bottle A, which must be perfectly clean and dry. Then fill the glass tube S, which holds exactly 10 cubic centimetres, with hydrochloric (muriatic) acid of the specific gravity 1·120. These 10 cubic centimetres absorb about 4 cubic centimetres of the carbonic acid, so that the weight of 4 cubic centimetres must be added to the result of each experiment.

The tubes C and D must be filled with water up to n, the top of D is then kept covered with a loose plug of cotton to exclude dust. While the water is being poured in, g must be kept open; g is then shut, the stopper carefully fitted into A, and the acid contained in S brought into contact with the pulverized limestone, as shown in the woodcut, by the tilting action of the hand. The carbonic acid evolved passes through r into B, and thence through the glass tube u into C. The water in C is forced downwards in proportion to the pressure of the gas upon it, and the water in D rises in a corresponding degree. Were the water in D allowed to remain higher than in C, the carbonic acid would be under a pressure exceeding that of the surrounding atmosphere. The water in D is therefore kept as nearly as possible at the same level as in C by opening the pinchcock p. If too much water has been allowed to escape through p, more may be introduced by blowing through v and opening p. In the same manner C and D may be refilled for the next experiment, only in this case g must be opened.

Before reading off the number of cubic centimetres, the apparatus should be allowed to stand sufficiently long to acquire the temperature of the surrounding atmosphere, the bottle A having been warmed by the hand, and the action of the acid on the substance ascertained.

Having measured the volume of carbonic acid, we can calculate its weight by means of the following formula:

$$\frac{C + 4 \times 0\cdot00197}{t \times (1 + 0\cdot0037)}$$

in which C indicates the number of cubic centimetres obtained, and t the temperature shown by the centigrade thermometer.

Let us suppose that 0·5 gramme of a calcareous substance has evolved 15 cubic centimetres of carbonic acid at a temperature of 12° C. The weight of this volume will be

$$\frac{(15 + 4) + 0\cdot00197}{12 + (1 + 0\cdot0037)}$$

equal 0·031076 gramme or 6·21 per cent. of carbonic acid, which has been combined with 7·89 per cent. of lime, so that the material thus examined contains 6·21 + 7·89, equal to 14·1 per cent. of carbonate of lime.

In the above calculation the barometric pressure has not been taken into account, and may in fact be disregarded for such practical purposes as those under discussion.

Should the substance analyzed contain caustic lime, this may be converted into carbonate by heating it with a solution of carbonate of ammonia, the excess of which may be driven off by gently warming it over a spirit lamp.

By means of Scheibler's apparatus, the estimation of the slip or slurry of the wet process, and the bricks or powder of the dry method, can be readily ascertained, and through its agency the accuracy of the raw cement mixture controlled in a few minutes. It is not necessary to calculate the percentage of carbonate of lime in every case. A sample of the raw material of known good quality may be taken as a standard, and a few experiments with the apparatus will soon show the operator to what degree the volume of carbonic acid may vary without affecting the quality of the cement.

Should the temperature vary, allowance must be made; the coefficient of expansion of carbonic acid being 0·0037 for each degree Celcius. If, however, the temperature at which each experiment is made is comparatively uniform, this correction may be omitted.

CHAPTER VII.

TREATMENT OF THE RAW MATERIALS.

WE cannot do better for the convenient illustration of the primary object of the treatment of carbonate of limes, silica, and alumina, than call attention to the experiments of Smeaton, as described in the first chapter of this book. We there see how he developed the nature of hydraulicity by estimating the amount of foreign matter by which it was occasioned. Our process will, however, be the reverse of that which was adopted in the Eddystone experiments, and our duty is to impart to the carbonate of lime the estimated quantity of the required ingredients. The calculation for guidance in estimating the exact quantities is discussed elsewhere, and we have only to consider here the most approved means for the accomplishment of this the very alphabet of Portland cement making.

There are three well-known operations, differing in character, the application of which is influenced or dictated by the quality of the materials used. They may be classed thus:

1st. Wet method,
2nd. Semi-wet method,
3rd. Dry method,

differing only in the mode, but having the same direct object of thoroughly amalgamating the different ingredients. They are all mechanical in character until the stage of burning is reached.

The wet process, from being the one most in use and generally dealing only with chalks and alluvial or other clays, is apparently the simplest in character and least costly in its first results. Chalk and clay, being capable of easy solubility, are charged in their progress through the "wash-mill" with three or four times their weight of water, and theoretically this process appears perfect, but practically, as we shall endeavour to show, there is considerable inconvenience, if not danger, even under the most careful supervision. The difference in the specific gravity of the chalk and clay involves an irregular deposition of the washed mixture, not only in its final settlement in the "back," but on its way along the shoots. In the most limited works this objection exists, and in the largest in a more aggravated degree, for the farther from the wash-mill the liquid material has to be propelled, the greater must be the amount of its irregular and eccentric precipitation. This part of the process may be likened to the river and tidal action, by which the water conveys the degraded portion of banks and shores in mechanical suspension, parting with each division of them in proportion to their fineness or texture, the finest going farthest. A familiar example of this action occurs in the river Thames, on the banks of which are deposited the fine particles in the form of mud, the mid-channel or stream receiving the heavier ones, such as sand and gravel. The finest of all in fact, the almost impalpable, are deposited at the point of stillness, when the tide checks the flow seawards. Near the river mouths or estuaries we find the most highly comminuted clay or mud.

Smeaton, by the chemical destruction of the matrix in which was deposited the silica and alumina, liberated it from its enthralment, and so unlocked the secret which he was desirous of knowing. The cement maker has to reduce his raw materials to such a degree of fineness as will permit of their accurate combination, and this can be effected in two ways—the one chemical, and the other mechanical. The former, the more expensive of the two, by converting the carbonate of lime into lime, and the silica into soluble silicates, and thus ensure an accurate result, but not one that a Portland cement maker can indulge in, and which we need not further discuss because of its costliness.

1st. The wet method consists in mixing the chalk and clay intimately, through the agency of water, in what is technically called a "wash-mill." This treatment is that which was adopted from the commencement of the cement manufacture, and varies in the manner in which the various triturating or reducing appliances are placed. In some works knives or cutters only are used, and in others to them are added harrow tines. In the early mills edge runners were also used in addition to the harrows. The following process is invariably one of decantation through the agency of the receptacles or back, into which the mixture is washed.

2nd. Semi-wet method, which was first adopted and patented by M. Dupont, in France, many years ago, passes the mixed chalk and clay between horizontal mills in a semi-plastic state. It is then in a fit condition to be placed on the drying floors, and the time usually occupied in waiting on the decanting of the water off the "slurry" in the "backs" is saved. Under the best conditions by which this process can be performed it must fall short in thorough mixing value of the wet process. This system has recently been introduced into this country under the name of "Goreham's patent process." Another adaptation is spoken of for the purpose of speedily eliminating the water of mixture by squeezing or pressing or some other equally "far-fetched" scheme. We shall in all probability next hear of a boiling down or roasting process, for it is singular how perverse the human mind is, and often revels in retrogressive action. If equal ingenuity and energy were applied to the study of reducing the water in the raw materials—for chalk contains from 20 to 30 per cent. of moisture, and clay as delivered by the barges from 50 to 60 per cent.—we should soon find a nearer approach to the following:

3rd, or dry method, which is simply the mechanical reduction of the raw materials by *suitable machinery*. When this is performed in a slovenly, careless, or ignorant manner, much danger and expense arise. Doubtless, this plan admits of no loose or "rule of thumb" treatment—chemically, the lines laid down must be implicitly adhered to, and mechanically, the reduction of the ingredients should be as nearly perfect as possible. You may err in having it too coarsely ground, but no objection can ever arise from the resulting powder being too fine. At least, on the score of subsequent advantage in the other parts of the process, the limit to the fineness is only measured by its cost. There is a danger attending this method which is sometimes a puzzle to the ignorant, and that is the want of proper care in applying the necessary moisture to render the mass plastic enough for moulding into bricks. If the water is carelessly thrown on, as is the case in "tempering" clay for brick making, there is a liability to separate the atoms of carbonate of lime from those of the silica and alumina, and thus destroy the necessary accuracy of combination so essential to an effective kiln result. The author had an amusing illustration of this effect while making a survey and reporting on a cement manufactory in North Wales. The factory was situated in the neighbourhood of lias materials, but the carboniferous limestone, also contiguous, was used and thus treated. It was first crushed in a stone breaker, from thence passed on to horizontal burr stones, when it was carelessly ground. The next operation was putting it into a wash-mill going at a velocity of about eight revolutions per minute, when it was mixed with a very fine clay. The difference in specific gravity of the two materials resulted in what the most ignorant could have foreseen, a separation, or at best a most imperfect combination, and to add to the risk attending such a clumsy operation, the mill had no arrangement for overflow or back room, and was stopped to take out the materials as they were *finished*. The consequence of this mode of *manufacture* resulted in a never-ceasing conflict with the users of the cement. To partially obviate the effects produced by this Welsh process, it was necessary to keep the cement for months in store until the free lime had been thoroughly hydrated to a point which reduced the danger of using it direct from the millstones. Some materials cannot be accurately blended together, even in their passage through the millstones, when they differ much in texture, and in such cases the better way to combine them is by a mixing machine. For instance, a crystalline stone in combination with a soft clay or shale would during the process of grinding be retarded by the interposition of the soft clay or shale between the crystals, and thus prevent its economical reduction.

CHAPTER VIII.

THE MANUFACTURE OF PORTLAND CEMENT FROM CHALK AND CLAY.

It is almost a work of supererogation to enter into any lengthened description of this well-known and simple process, but as the task we have set ourselves to do is intended to describe every known system from the best sources of information at our command, we will proceed to discuss the treatment of chalk and clay.

Where large and well-circumstanced works exist, they will be found erected on the chalk formation. To this point the river clay is to be brought from the various sources of its deposition, and mixed with the chalk. The *modus operandi* is clumsy and careless.

The wash-mills (described at page 207) are erected at convenient points, and the chalk and clay are tipped in the required proportions, always a slightly varying one, for the contained moisture may be regarded as an unknown or doubtfully ascertained quantity. In works where probably close upon a thousand tons of chalk and clay are passed through per day, active and careful testing can adjust the average error from the wash-mill, leaving only the danger to be apprehended from the irregular deposition of the washed materials in the backs or reservoirs. If it be conceded that the cost of liquid or wet reduction of the raw materials is less than that incurred by the dry process, it must not be lost sight of that the former requires much more attention in remedying the results of its defects. The wash-mills require frequent cleaning out, the shoots or conduits must be kept clear, both of which involve a considerable percentage of waste. Then again the "luting" or mixing the liquid materials by agitation and the subsequent attention to the process of decantation of the superfluous water. Even with all this attention and expense you have to wait probably for months, until the contents of the "back" are in a fit condition for removal to the "drying plates," from which the dried material is taken to the kilns in a state much inferior for chemical or kiln treatment to the brick pressed from the powder of the dry method. It is impossible to make an accurate comparative statement of the relative cost of the two systems, owing to the fluctuating value of the quality of the materials dealt with by the dry method; the material from the lias formation costing less in its reduction than those from the more crystalline carboniferous, or oolitic formations. But we may without the least hesitation assert that, with the most improved reducing machines, the product from the dry process can be delivered into the kiln at a considerably less cost—having due regard to all the disadvantages of the necessarily widely separated backs and kilns—without taking credit for the saving in time. A strong way of putting it would be simply to state that while the back was being filled by the one method, you can by the other fill the kiln not only ton for ton (dried material of course), but three to one; and further, before the contents of the back are ready for final treatment, the cement from the direct process has been sold or used some weeks and perhaps months. Let us examine carefully the conditions inseparable from the washing process.

First, there is the latent moisture of the chalk itself, amounting on the average of all weathers to 25 per cent., and the clay to nearly 60 per cent. So that really the ton of chalk is only 15 cwt., and the ton of clay 8 cwt. The quantity of water applied to the mixture in the wash-mill is generally on an average 4 tons to 1 ton of the raw materials. The cost of conducting it by gravitation to the settling back is not great, but in some manufactories it is pumped to a higher level. It is not contended in this examination that the materials used in the dry method are free from natural or latent moisture, for in fact the hardest of all materials, granite, contains something like 12 per cent., but the preliminary process is the very converse of the wet, for instead of adding water, the first duty is to extract it. In the succeeding process, namely, that of converting the powder into a condition which will ensure its fitness for the kiln, we hope, before concluding our arguments, to show that a small percentage of moisture only will be required to effect that result.

The contents of the "backs" are not when first filled always accurately blended, and require careful testing from different points to ensure that the mixture, so heavily loaded with water, is true enough to be allowed to settle down.

If we grant that in the wash-mill the mixture is as perfect as such a process admits of, there is great liability of error before final settlement. Testing the raw product is therefore attended with difficulty and uncertainty, for you may be satisfied that on emerging from the mill or the conduit it is truly proportioned, but there is still ahead the irregular settling down in the reservoirs; so that practically your security is illusory, until the actual adjustment of the atoms or particles into a cohesive condition has been reached. Another series of testing is then obligatory, and to be useful must be of a more extensive character than that of the simple flow of the products from the mill. The samples must range over the surface of the back and reach also its various depths. These are precautions which no sensible cement maker can now afford to neglect, unless he desires to live in a continual state of alarm and anxiety.

In past years, in what might be called the dawn before the transition state of the trade, many of the Thames cement makers led the reverse of contented or happy lives. It is to be hoped that all of them have now bid final adieu to the "Fools' Paradise," in which they had occasionally to pass fitful periods of probation as penance for their sins of omission committed at the "backs"; or, in other and more appropriate language, "careless or negligent testing of the raw materials."

In supplying the wash-mill with the chalk and clay a measure of capacity is of necessity adopted. There being a constant variation in the condition of these materials, any measure of weight is out of the question. Under such circumstances the operator charged with the duty of supplying the wash-mill trusts entirely to the tally or marking system, and probably even its records are of an uncertain and treacherous character. The dry system admits of weighing accurately and with the materials in such a condition as to ensure evenness of result. In such a state the material offers unusually favourable facilities for testing, which through the agency of the apparatus described at page 154 can be performed in a few minutes, and thus ensure that not more than half a ton at most of the raw powder can at any time pass to the forming machine in a state of disproportion or inaccuracy. Such facility for challenging the faulty mixture is not commanded by the wet process, but on the contrary has the disadvantage, amongst many others, of involving delay in its ultimate realization. Entire trust has to be reposed in the manipulation of the man at the wash-mill, and his inattention or inexperience may and does in the best managed works lead occasionally to a "back" of almost worthless material, which the most expert cement maker does not always succeed in utilizing. The facility of having under your own control and within the space of a few yards from the kiln the raw material is one of great importance, and cannot fail eventually to supersede a system where in large factories the raw material is sometimes so remote in distance as a mile from the machinery of ultimate conversion.

The wet process of cement making may be compared to the bleaching trade in Lancashire before the discovery of chloride of lime. To carry on the trade previous to that period, extensive acreage was required to provide for the necessary surface while waiting on the desired result produced by the air or weather system. Chemistry, with its magic wand, transformed the acres into yards and the weeks into hours. Cotton manufacturers are now doubling the steam pressure in their boilers, so as to increase the speed of their spindles. Like the "great Napoleon," they disregard those racing before them (believing that they themselves are first), but have an eye to those pressing on them from behind. There are America and India close at their heels ready to take advantage of any false step.

The Stettin cement manufactories, placed on the banks of the Oder, are very extensive, and have the command of first-rate clay and chalk. One of the factories has a supply of both these materials on the premises. Fuel of the necessary quality is of course more expensive than in England, and much of the coke hitherto used has been produced from English coal. The introduction of the Hoffman or ring kiln, however, gets over much of this difficulty, as through its agency a very ordinary quality of coal can be used.

There have been various methods used for the amalgamation of the chalk and clay. One (which is probably now abandoned) was first to wash the chalk into a back, and after a certain period had elapsed, the value of the contents of the back were accurately ascertained—that is, the value in their dried state. The clay was then finely powdered and mixed with the washed chalk in the cement proportions on its passage through the brick-forming machine, from which the bricks passed of the required size of $9'' \times 4\frac{1}{2}'' \times 2\frac{1}{4}''$.

In another manufactory the two materials were originally mixed together in a species of mortar-mill, with two edge-runners. After a certain time of settlement, in unusually small backs, the slurry was taken out in rather a wet condition, and was mixed with dried material, so as to enable the mixture to be passed through a brick-making machine and formed into bricks of the usual size.

These manufactories labour under a considerable disadvantage from their being absolutely closed during two months in the severest time of the winter, when no labour can be carried on.

CHAPTER IX.

THE MANUFACTURE OF PORTLAND CEMENT FROM THE BLUE LIAS MATERIALS.

OWING to the imperfect and careless manner formerly adopted in the blue lias districts, very unsatisfactory products were placed in the market, having a most damaging effect on the reputation of such cements. Before and during the year 1868, the author in common with most engineers looked upon "Portland cement" from Warwickshire and Somersetshire (the more important seats of such manufacture) with much suspicion, and generally characterized it as worthless and false. Its sale was limited and confined to small consumers, to whom its best recommendation was cheapness, and within so circumscribed a limit of intelligence the damage sustained was not of a very serious character. In the early part of the year 1869 the author was consulted as to the best method of extricating the Rugby cement works from a dead-lock caused by misunderstanding in the management. On visiting these works he found that raw materials of a valuable character were in abundance, but a total absence of the necessary technical skill for their proper and profitable conversion. There was *no Portland cement* made there at that time, and the following prescribed formula will show that its production under such conditions was impossible.

For the manufacture of best Portland the prescription was:

In Winter.
Stone .. .. 5 parts.
Clay .. .. 4 „

In Summer.
Stone .. .. 1 part.
Clay .. .. 1 „

These proportions were roughly mingled by the workman as they came from the quarry.

There was also produced what was termed light Portland cement by the following mixtures:

In Summer.

	Burnt Shale.	Best Portland.	Light Brick.	Foundry Coke.
Parts	6	2	2	3

In Winter.

"	6	2	3	3

And for heavy natural cement:

	Heavy Natural Stone.	Burnt Shale.	Best Portland.	Light Brick.
Parts	9	3	2	2

Lias cement again was made of:

	Burnt Shale.	Best Portland.	Light Brick.
Parts	3	1	1

while "Roman cement" was produced from:

	Burnt Shale.	Colouring Stone.	Light Brick.
Parts	3	½	½

The manipulation of these several materials to realize the various results was performed in the most careless manner, and under no special measures of capacity except the ordinary and irregular sized wagons of about one cubic yard in volume used for general purposes at the works.

We will explain a little more fully the nature of these several materials.

"Stone."—Containing about 80 per cent. carbonate of lime.

"Clay."—Disintegrated shale, generally containing about 20 per cent. carbonate of lime, about 60 per cent. silica, and 10 per cent. alumina.

"Shale."—One of the beds of indurated shale, containing about the same proportions as the clay.

"Best Portland."—Clinker taken from the kiln when the best Portland cement was burnt.

"Light brick."—The yellow or slightly burnt bricks from the same source as the last mentioned.

"Foundry coke"—The best coke, as commonly used in the cupola of an iron foundry.

"Heavy natural stone."—A bed of stone approaching in chemical value to the proportions required to make a Portland cement.

"Burnt shale."—Shale specially burned for the purpose of these mixtures.

"Colouring stone."—An imported material, obtained from the deposits of argillaceous ironstone in the neighbouring Northamptonshire oolitic deposits.

Such a jumble of mixtures and proportions clearly indicated that the management had no reliable knowledge of the true constituents of a good and honest Portland cement. The author had some difficulty, even after he was invested by the partners of the works with full authority, in upsetting this ignorant and damaging process of manufacture, and directed his attention to the more simple process of making a good Portland cement. The materials were carefully examined and analyzed as well as the circumstances of the works permitted, for it should be mentioned that these were originally used for making bricks from the clays and shales. It was not long before a good heavy cement was produced by the following process:

The *raw* materials (stone and best shales) were accurately mixed in their rough condition, passed through toothed rollers, and thence to the horizontal millstones, from the spouts of which the finely ground powder was elevated to the dust-room. The next part of the process was that of passing the dust by double worms or creepers descending a shoot, where it received the necessary amount of water to render it sufficiently plastic for the pug-mill brick-making machine, through the dies of which it rolled on to trays which were passed into flues constructed on " Beart's patent principle." After being sufficiently desiccated, the bricks were pushed through these ovens and burnt in the old brick-kiln without cover of any kind. The kilns were oblong in form, and about 10 or 12 feet high. The bricks were built in layers, having placed between each course (as is done in other kilns) the necessary quantity of fuel in the shape of gas coke. The bricks were well clinkered, and, much to the author's satisfaction, produced really a good heavy cement. The kilns had underneath the common "dead horses" or arches, which were used to light it up. These kilns were unusually extravagant in the cost of fuel, but owing to the difficulty attending the building of new and appropriate ones, their service was continued, and, considering the circumstances, with much success.

As may well be supposed, works producing (under the old system) such a medley of cements, created amongst the customers who were confiding enough to purchase its products much eventual dissatisfaction and distrust. Under such circumstances the author pressed on the market of Lancashire a *good* Portland cement, the acceptance of which was difficult, but owing to confidence in the writer as an acknowledged authority on Portland cement, and to his selling conditionally that if not good no charge was to be made, he eventually succeeded. It was necessary also to sell, for some time at least, this good cement at the same price as was formerly charged for the previously prepared rubbish.

In addition to the difficulties of such damaging reputation, the author had to contend with a large list of old claims for damages, which were ultimately met.

The quality proved to be as good as represented, and the reputation of cement made from blue lias materials was established. The cement continues to maintain its reputation, and can hold its own against the best-known London cement, unless where ignorance and prejudice stand in the way.

These remarks are perhaps of greater length than the subject may appear to warrant, but the author is desirous that a cement-making source of so valuable a character—especially in Warwickshire—should not suffer from the misrepresentation of its interested opponents. The products of the Warwickshire cement works find their natural outlet in a northern direction, for the cost of carriage to London operates against them entering into any large successful competition with the Thames and Medway cements. In the northern counties, however, a large and increasing trade is done, and with a daily improving reputation, proving that in one inland district at least a good Portland cement can be made, equal in every respect to the best and oldest established London cements.

In the production of Portland cement from these materials there is one great advantage they possess over almost every other—from whatever geological formation—that is, the small and almost inappreciable risk from cracking. The carbonate of lime s in so finely comminuted a state, and so accurately blended with the silica and alumina, that no injurious development from this source can possibly arise, at all events in the direction of the cracking or blowing danger.

The facility with which cements of a fast-setting type can be made from the lias materials is too tempting when a market can be found in which they may be consumed. The character of the whole of the lias localities has suffered in consequence, and it will still be a long time before it can be finally recovered. Notwithstanding the enlarged experience and improved knowledge of the subject, these imperfect cements are unfortunately still prepared with such results as that described in Chapter XX.

The existence of all the required raw materials for the preparation of good Portland cement, and the facility with which they can be converted, will ere long stimulate the manufacturers in the lias districts to a more safe and truthful development of their undeniably great advantages. There are lias deposits on some of our seaboards from whence there could be easily nurtured a large and profitable export or foreign trade. There can be no question about the various facilities the lias district possesses over any other analogous geological deposits, and the more important is the readiness and shortness of time in which a cement can be converted from the limestones and shales. In all weathers and on the same ground from the time of quarrying to the filling of the sack or cask for the market, one week is ample for the process of manufacture, and the improved machinery and methods of conversion by improved kilns is likely before long to reduce even that short period. While therefore the wet or chalk and alluvial clay converters are waiting on the protracted results from their manipulation, a dry or lias, and other similarly circumstanced manufacturer can turn his capital round at least a dozen times.

But the Thames cement makers insist and proclaim that it is not Portland cement if made from any other material than the London chalk and clay. They forget that the inventor of the name (Aspdin) called his material Portland cement from its resemblance to the Portland stone, and it was obtained from materials other than London chalk or alluvial clay. If Aspdin or those with whom he had been associated possessed or could have commanded the required chemical knowledge, they never would have gone south to operate on materials whose best recommendation besides their great purity was the facility with which they could be rendered soluble. This circumstance led to the institution of a clumsy and hap-hazard system of manufacture, so that in 1859, when called upon by the Metropolitan Board for a reliable cement that would stand a tensile strain of less than 200 lbs. to the square inch, not more than 20 per cent. of the manufacturers could accept or perform such conditions. They at that time received a lesson which has proved beneficial to themselves and their customers. Instead of boasting, however, of the acres covered by their works and of the miles of tramways by which they are traversed, it would be better for their eventual reputation to consider whether the time has not arrived to reduce the surface of the one and the ramifications of the other, and thereby bring down the cost of their productions.

There is much difficulty in removing prejudices, and probably it will take as long to establish an improvement in making Portland cement as it has done to introduce machinery in the cognate process of brick making. That art fifty years ago occupied about the same relationship to the improved machinery of to-day, that the Thames and Medway cement works now do to the improved mechanical and chemical system commonly known by the name of the dry process. It is not long since some of the implements used in brick making were identical with those employed by the ancient Egyptians in making their bricks. The same form of Portland cement kilns in which Aspdin burnt his materials are with but little difference the kilns of to-day.

The increasing magnitude of the Thames and Medway cement works raises a question of their being quite correctly placed in the midst of or contiguous to densely populated districts. It is just possible that such manufactories as now conducted may be controlled and regulated under the provisions of the Noxious Vapours Act.

There is not now much remaining for us to say regarding the production of Portland cement from lias materials, except to remark that instead of a measure of capacity one of weight should be substituted, and it would be more advisable to pass the stones and shales through the breaker before performing the mixing operation. By this means you would the more accurately secure the homogeneity of the raw mass, and prepare it for its perfect amalgamation in its passage through the horizontal millstones.

In dealing with the dust, a brick-making machine should be selected capable of compressing the mass into as solid a brick as possible. The more pressure you can apply the better, as you are enabled in that case to reduce the quantity of water, and thereby express from the brick as much of the contained air as possible. The particles cannot be brought into too close contact if the best results are to be secured from the action of the heat in the kiln. In a lias quarry there will always be found a large preponderance of shales, and as some of these contain an undue or rather an unsafe quantity of iron pyrites, it requires care to select the purest. The existence of the shale difficulty is doubtless the reason why past as well as present cement makers in the lias districts make such a variety of cements, and thereby utilize the waste.

The burning of the cement in open kilns does not really interfere with the value of the cement, but, on the contrary, facilitates the expulsion of the carbonic acid gas. Another advantage which is also realized by the ring or Hoffman type of kiln, is the readiness with which the kilns can be loaded and unloaded, being on the same level as the general floor of the manufactory. One of the objections, as before observed, is that of extra cost of fuel, which the open and exposed form renders unavoidable. There is, unless when the conditions of the weather are favourable, a sluggish and uncertain combustion, which, in the absence of chimney or shaft, provides no means of regulating or increasing the draught. Although operating in an extemporized cement works, and generally speaking in the absence of special advantages which would otherwise be secured, the author found that a good and reliable heavy cement could be made at less cost of fuel than is required to produce a cement of equal weight in the chalk districts. There is not much doubt that in times of ordinary cheap fuel and labour Portland cement could be produced, including every cost, interest on capital, and all contingencies, at eighteen shillings per ton. To accomplish this, however, it would be necessary that the manufactory should be erected with special reference to the duty required of it, and command the most recently improved reducing machines, as well as the best form of kiln.

This is perhaps as good a place as any to examine what is required in the reduction of the raw materials, so as to fit them for the brick-forming operation. It will be necessary to take the most careful means to ensure that the powder is even in quality or texture, and that none of the grains are larger than can be passed through a sieve of 900 meshes to the square inch. The dry process has not at this stage the advantage commanded by the washing or wet one. Such materials, capable of being dissolved by the wash-mill, are limited to chalk, marls, clay, with others of comparative insignificance, and when, as they frequently are, alloyed with insoluble ingredients, the washing operation fails to purge them unless by the exercise of the strictest precautions.

CHAPTER X.

THE MANUFACTURE OF PORTLAND CEMENT FROM THE CARBONIFEROUS AND OTHER LIMESTONES.

THE wide range of quality in the various limestones, whether we regard their physical character or their value in a chemical direction, is very considerable. Generally speaking they may be considered as too hard or crystalline in texture for such treatment as that bestowed on the chalks and other amorphous materials. This obduracy of natural constitution necessitates the application of effective and specially adapted machinery for its thorough, and as nearly as possible perfect, reduction. There can be no question that the manufacturer using such materials is burdened at the initial stage of his operations with expensive machinery not required by those engaged in the conversion of chalks and clays. These unkindly qualities of this class of materials have led to many suggestions for their compound and other treatment. The double-kilning process, as it is sometimes called, destroys the crystals, and in fact the whole structure of the stone by its conversion into lime. It is then properly quenched or slaked, and reduced to a pasty condition, and in this hydrated form is mixed with the clay sometimes in a finely powdered state, and also as a semi-fluid. The preferable method would be to introduce the powdered clay while the mass is passing through the brick-moulding stage, not directly into the pug-mill proper of the machine, but after it had traversed a properly defined space, provided with double screws, working on two spindles or shafts going in a parallel direction. This would ensure the homogeneity of the mass and secure its perfect amalgamation in the brick machine. One great advantage in this treatment is the eventual saving in the fuel required for the kiln. There is no doubt, however, that, although securing the most perfect condition for conversion, it involves an extra cost in the initial process of decarbonization of the limestone. Such a cost, however, depends on the locality in which the process is performed, and must be influenced entirely by the nearness or remoteness of the fuel. Some makers have endeavoured to substitute a partial process of calcination, but, so far as the author knows, with but little success. If you adopt this system, either go the whole length or fall back on the reducing machinery, no middle course can be dealt in with safety.

The materials having been duly selected and their chemical value accurately ascertained, we have only to adopt the best devised machine for their perfect reduction. In Chapter XI. the various machines are minutely described, but we may here shortly follow the various processes.

The Blake's crusher will in all cases be found the machine best adapted for the treatment of rough materials, and from which they pass into the secondary or Goodman's machine. At this stage they are susceptible of easy mixture with each other, for our opinion is in favour of separate and distinct treatment until they reach the horizontal millstones to be finished. Much will depend on the character of the clays and shales with which the manufacturer has to operate.

A short description of a compound treatment as successfully adopted at Doveholes, in Derbyshire, will perhaps be the best illustration of what is not by any means a very difficult process.

The materials dealt with are lime waste, shale, and carboniferous limestone.

The waste is first thoroughly desiccated, becoming by that treatment a powdery and easily reduced mass. The shale, owing to the present supply being of a heterogeneous character, and in some of its component parts impregnated with iron pyrites, is mildly burnt in large heaps in a manner somewhat similar to that adopted in converting clay into railway ballast. The limestone is reduced to pieces the size of hazel nuts by the agency of a Blake and Goodman crusher. All these several materials are then in a fit condition to be taken to the horizontal millstones, through which, in their proper proportions, they pass on at once in a fine powdery state direct to the brick-making machine, through which at present they are pugged, regardless of their form, and conveyed to the drying plates for final desiccation. In irregular lumps it is in a few hours passed on to the kilns and laid in alternate layers with good hard coke. About 75 per cent. of the lime element in this mixture being hydrate of lime, a considerable saving in fuel is thereby effected. Even with this favourably circumstanced mixture some danger is risked from the liability of the particles to get separated from their desired closeness of contact by the irregular running of the millstones, and also on the application of the water of plasticity, which if carelessly or irregularly distributed leads to a faulty mass, which in the kiln results in a *dusting* or imperfect clinker. In any new arrangement for the conversion of raw materials, differing widely in texture and specific gravity, it would be preferable to provide for the separate grinding of each of the materials, and afterwards mix them by a double hopper before entering the carrying screws to the brick-mill. The proportions could thus be more accurately blended and the kiln results thereby made more even and regular. No trouble should be spared in making the material at this stage as perfect as possible, for any carelessness here will entail an amount of loss in the concluding processes, which no subsequent provision can remedy.

There is an advantage in having a large quantity of raw powder in advance of your wants—say at least twenty-four hours' supply, for a period of soaking of that duration has a most beneficial influence on the quality of the resulting cement. In adopting this method there would be an extra cost in moving the mass so retained to the brick machine. The improvement of the cement, however, more than counterbalances any question of extra labour.

The dry method may, even with such progress as has been made, be regarded as in its infancy. The earlier experimenters in this country, at all events, have been generally ignorant of the requisite chemical knowledge and too narrow-minded or impatient to call to their aid the assistance of specially skilled experience. Some, without the suitable machinery, have failed, because they could not succeed in reducing the materials fine enough to ensure successful results. Notwithstanding the extent of these disappointments, they are as nothing compared to the mistakes and blunders of the early operators in chalks and clays, who stumbled in the dark in the midst of great and persistent opposition on the part of the Roman cement makers.

CHAPTER XI.

MACHINERY OF REDUCTION.

The kind of machinery required for this purpose will greatly depend on the system pursued. By the wet method the wash-mill treats the raw materials so as to fit them for the succeeding operation of drying and burning. The dry method requires the assistance of stone-breakers, triturators, or other reducers, before the raw materials can be rendered suitable for the processes of desiccating or burning. From this point the machine treatment of the resulting clinker is common to both systems, and their description or a discussion of their respective merits will be equally applicable to the raw as well as the calcined materials.

The most widely known class of stone-breaking machines are those of the Blake type, and their general adoption in nearly all parts of the world sufficiently indicates that the leading principle of their construction is well suited for the purposes to which they are generally applied. The numerous failures that have arisen during the last twenty years in the search after a reliable stone-breaker indicated that such a machine was a necessity, and there is some satisfaction in knowing that the various and gradual improvements on the original "Blake" now almost reach the point of perfection attainable through that design.

So much depends on the successful and economical reduction of the materials from which Portland cement is made, that we regard a full and searching examination of the machinery through whose agency it is effected as highly necessary. The most ready way perhaps to give some idea of its importance, and to illustrate the various stages of this machine's development, will be by tracing its several modifications from, as nearly as we can ascertain, its original source. There are two leading features in all the machines, of whatever form they may be, which indeed have led to its success. The one the eccentric motion from the shaft, and the other the movable jaw action, and without which a Blake's stone-breaker would be almost as useless as a cart without wheels. The act of breaking stones, even with the hammer intelligently guided by the hand, is not only laborious, but costly. A machine designed to imitate this action, such as stampers actuated by gearing and cranks coming as they did into direct and violent contact with the materials under treatment, developed an amount of concussion destructive in its character and fruitless in its effects, frequently resulting in damage to the machine, however strongly made. The idea therefore would naturally occur to the machinist to obviate this objectionable effect by a distribution of the shock through the skilful combination of the different parts of the machine. Such necessary design must have occurred to the original contriver of the machine, and we take from an American book an illustration of stone-breakers at work in the Central Park, New York, probably about twenty years ago. Fig. 3 is a longitudinal section of the essential parts of a stone-breaking machine, A A' is the frame of cast-iron in a single piece, which receives and supports the other parts. This frame consists of two parallel cheeks, A, connected together by the parts A' A (shaded with diagonal lines). The arc B represents a fly-wheel, of which there are two, one on each side of the frame, working on a shaft having its bearing on the frame. This shaft is formed into a *crank*, E, between the bearings, and carries a *pulley*, C, to receive a belt from a steam engine or other driver. The fly-wheel, the section of fly-wheel shaft, the pulley, and the arc described by the centre of the crank in its revolution, are indicated by dotted circles. F is

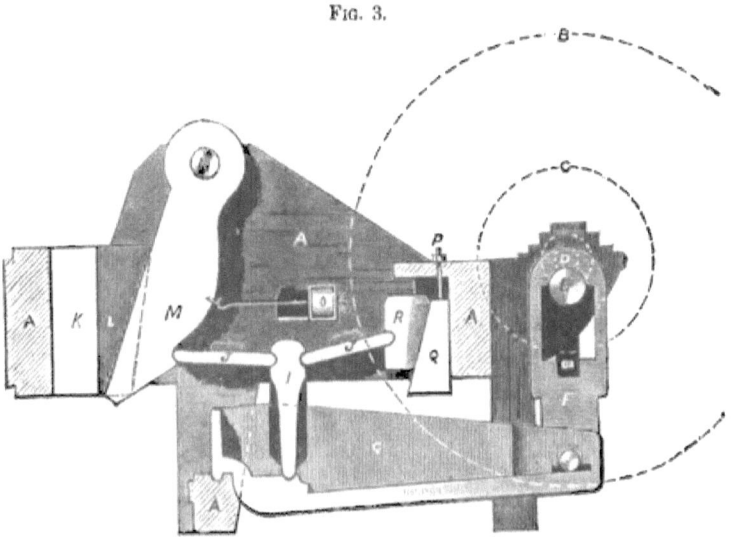

Fig. 3.

a *pitman*, or rod which connects the crank with the *lever* G. This lever has its fulcrum on the frame at H. A *vertical* piece, I, stands upon the lever, against the top of which piece the *toggles* J J have their bearings, forming an elbow or toggle-joint. K is the *fixed jaw*, against which the stones are crushed. This is bedded in zinc against the end of the frame, and held back to its place by checks, L, that fit in recesses in the interior of the frame on each side. M is the *movable jaw*. This is supported by the round bar of iron, N, which passes freely through it, and forms the pivot on which it vibrates. O is a spring of indiarubber, which is compressed by the forward movement of the jaw, and aids its return.

Every revolution of the crank causes the lower end of the movable jaw to advance towards the fixed jaw about a quarter of an inch and return. Hence if a stone be dropped in between the convergent faces of the jaw it will be broken by the next succeeding bite; the resulting fragments will then fall lower down and be broken again, and so on, until they are made small enough to pass out at the bottom. The distance between the jaws at the bottom limits the size of the fragments, and may be regulated at pleasure. A variation to the extent of five-eighths of an inch may be made by turning the screw-nut P, which raises or lowers the *wedge* Q, and moves the *toggle-block* R forward or back. Further variations may be made by substituting for the toggles J J, or either of them, others that are longer or shorter; entire toggles of different lengths being furnished for this purpose.

The proper speed for this machine was 200 revolutions per minute.

It will be observed that the New York stone-breaker had a crank instead of an eccentric, and although it may appear complicated in comparison with the improved modern machine, there would be, although perhaps obtained by increase of friction and multiplicity of parts, an almost entire freedom from shock or concussion. The most modern form of this description of stone-breaker has not lost much of its original character, for its two leading characteristics in all its modified forms are the jaw and the crank (eccentric) action. It does not appear that the above-described breaker had acquired at the time mentioned any distinctive name, but we may fairly assume that it was the machine from which all the others have sprung, and clearly illustrates with what tenacity every succeeding improver has clung to its two salient features or points of excellence.

We will now proceed to describe the various improved patterns of this machine made since its first introduction into this country. It will be found that the numerous altered designs indicate a strong desire to mitigate the vibrations of the shock while working, and to reduce as much as possible the wearing parts. The general appreciation of Blake's machine indicates that in principle it is correct, for the strong hold it has in public estimation favours such a supposition.

It was invented by Blake, an American, and exhibited in London in 1862. It was about that time patented by Marsden, who until 1875 possessed sole right to manufacture the stone-breaker, and exercised his privilege so exclusively that all of these machines until that time were made by himself or under his supervision and control. Such a monopoly so rigidly exercised shut out of the market all attempted improvements (many of which resulted in heavy and expensive litigation in their defence), and encouraged, or rather tolerated, a careless make of the machine itself. Since the expiry of the patent several improvements have been made, but they are of such recent introduction that very little opportunity has been afforded of testing their respective merits. Fig. 4 shows a Blake's stone-breaker.

The primary object in the Blake's machine is the obtainment of the crushing force or power from an eccentric action of the shaft D, which is driven by a pulley fixed on one of the fly-wheels. There is a fly-wheel, B, on either side of this shaft, and the speed is usually about 250 revolutions per minute. From the shaft is depended a connecting rod, E, at the bottom of which are fitted movable toggle-plates, G G, on either side. The eccentric action of the shaft at each revolution imparts a swinging or lateral motion of the connecting rod, and through the toggle plates it is conveyed to the jaw, J. An indiarubber spring attached to the jaw at its lower extremity draws it back after completion of the stroke. The action is simple in character, and under favourable conditions these machines produce excellent results at a comparatively low expenditure of power. They are easily adjusted to suit the varying sizes of the materials required by changing the toggle-plates.

The comparatively acute angle also with which the movable jaw impinges against the fixed one induces a tendency to

FIG. 4.

slip the stones out, and accidents have occasionally arisen from this cause. It will be found that the most economical results are obtained when the materials operated upon are free from other than latent moisture, and as they approach a crystalline character the greater will be the output from the machine. The fracture of the stones caused by the stroke of the movable jaw necessarily produces a considerable amount of dust, which, from a cement-making point of view, is not objectionable, but, on the contrary, beneficial.

The Blake machine is only here recommended as the first agent for reducing the larger materials to a size about four cubic inches to supply an intermediate reducer before the passage of the materials to the final operation performed by the millstones. A Blake's machine worked under such conditions would prove a most valuable auxiliary in cement making, instead of occupying the ordinary and misunderstood position which it now generally does. The Blake's machine is not designed, nor by its mechanical arrangement is it adapted, for the fine reduction of minerals or materials of any kind. When such work is sought from it nothing but disappointment arises, owing to its unfitness for other than the work for which it is specially made. This machine should receive from the quarrymen the stones on which it is to operate, and pass them broken and crushed in sizes not less than three or more than four cubic inches. This produce should then be further reduced by a machine of a different character, the leading features of which will be hereafter described.

There have been many attempted improvements on the original "Blake," but as these more or less pirated the particular form of jaw upon which Marsden's patent was supposed to be founded, their manufacture and use was prohibited by legal injunction. Thus baffled in the adoption of the eccentric and jaw action, various inventors devised machines of a clumsy and inefficient character, whose merits were not striking enough to obtain for them any lasting or definite position as useful or economical stone-breakers. The expiry of Marsden's patent opened the field for legitimate improvements, and we shall now proceed to point out their nature and character. The two recently patented improvements on the "Blake" are perhaps entitled to our first consideration, from the fact of their emanating from gentlemen who had been for a long time identified with Marsden in the make and commercial development of his machine.

Hall's "multiple action" machine, as shown in Figs. 5, 6, 7, is pretty nearly in form and pattern like the original. The

Fig. 5.

inventor of this improvement claims an advantage in effecting a considerable reduction in its weight, owing to the avoidance, through his peculiar disposal, of the jaw action, by which all strain and consequent vibration is avoided. Another advantage is also gained by Mr. Hall in substituting for the troublesome and unreliable indiarubber spring a draw-back motion self-regulating and self-acting. Fig. 5 shows the position of this arrangement, X being the coupling rods connected by a cross lever Y mounted on a stud at the back of the machine; the jaws securing by this arrangement their accurate withdrawal after the stroke has been performed.

Fig. 6.

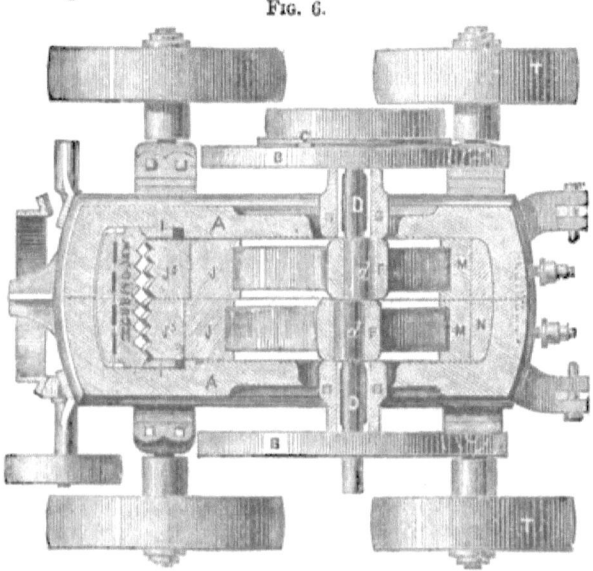

Fig. 7.

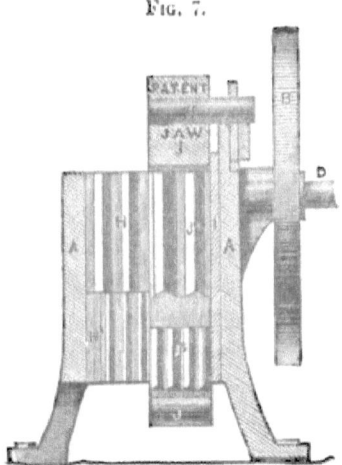

The primary feature, and indeed the salient point of this improvement, consists in dividing the jaws as shown in

Fig. 6, the driving shaft, D, having two eccentrics, d, d, each of which actuates its respective part of the divided jaws, J, J, so that one is receding while the other is advancing. In fact we may almost describe this as a double machine. The advantage which appears to us obtainable through this combination of jaws, apart from the merits attributed to it by the inventor, is the prevention of jamming, an important advantage, for much time is frequently lost in freeing the stones when they get wedged and beyond the impinging influence of the jaws. Accidents are frequent from this cause, owing to the workman hammering down the stones into more intimate contact with the jaws. The new multiple arrangement will tend to overcome this difficulty, for there will be a constant vibratory action facilitating the passage of such stones as either from their configuration or condition have a tendency to choke or obstruct the machine. Sometimes this derangement arises from careless feeding of the machine, or the introduction of stones too large to be profitably acted upon.

Mr. Hall further meets a want by introducing a system of jaw reparation, as shown in Fig. 7. Those familiar with the use of stone-breakers have experienced the costly character of the jaws, which so frequently require renewal. The greatest amount of wear is at the lower end of the jaw, which being hitherto in one piece, considerable waste was incurred in having to abandon jaws only partially worn.

The moving jaws J (Fig. 7) have a raised projection J^1, which is chilled on the top, cast across the face; this is undercut at each side and suitable faces let in, and held at the top at J^2 and bottom J^3 by wedges or wedge-shaped bolts, which can be easily tightened without the removal of any parts. The fixed jaws are also made in two pieces, and each particular section of face can be reversed, so as to present two wearing surfaces.

There are in Mr. Hall's machine three marked and significant alterations claimed by him as of importance and advantage.

1st. As most beneficial, the division of the hammer or striking jaw.

2nd. The arrangements for ensuring the accurate return of the jaw after performing the stroke of impingement.

3rd. The facility with which portions of the wearing jaw faces can be removed and fresh ones substituted.

From a practical point of view, the second and third alterations are most desirable, and should result in much economy. The division of the hammer-jaw is of course attended with an increase of friction and a liability to wear from being exposed to the dust at the joints. Those accustomed to break sub-crystalline limestone with the old Blake's machine will be familiar with the large quantity of almost impalpable dust produced, and it is from this source that liability to derangement and wear and tear are to be apprehended. The improver of this machine has probably not overlooked this, what the writer regards as a moderate weakness, and has prepared a remedy.

Messrs. Broadbent's improved Blake stone-breaker consists of an altered design of the machine, having a special arrangement called a "patent positive draw-back motion," as shown in Fig. 8. There can be no doubt that this is a great improvement, for the original indiarubber spring was unreliable, and when the accuracy of its adjustment was neglected much waste of power and consequent inefficiency of results took place. A reliable arrangement of this kind ensures the return of the jaw, and consequently will much diminish, if not entirely overcome, the tendency to fouling or choking of the machine. Hitherto in all the old and original types of the Blake the impulsive action was primarily imparted to the jaw through the eccentric of the shaft, and the

MACHINERY OF REDUCTION. 193

provision for the rebound was entirely dependent on the elasticity and accurate adjustment of the indiarubber washer

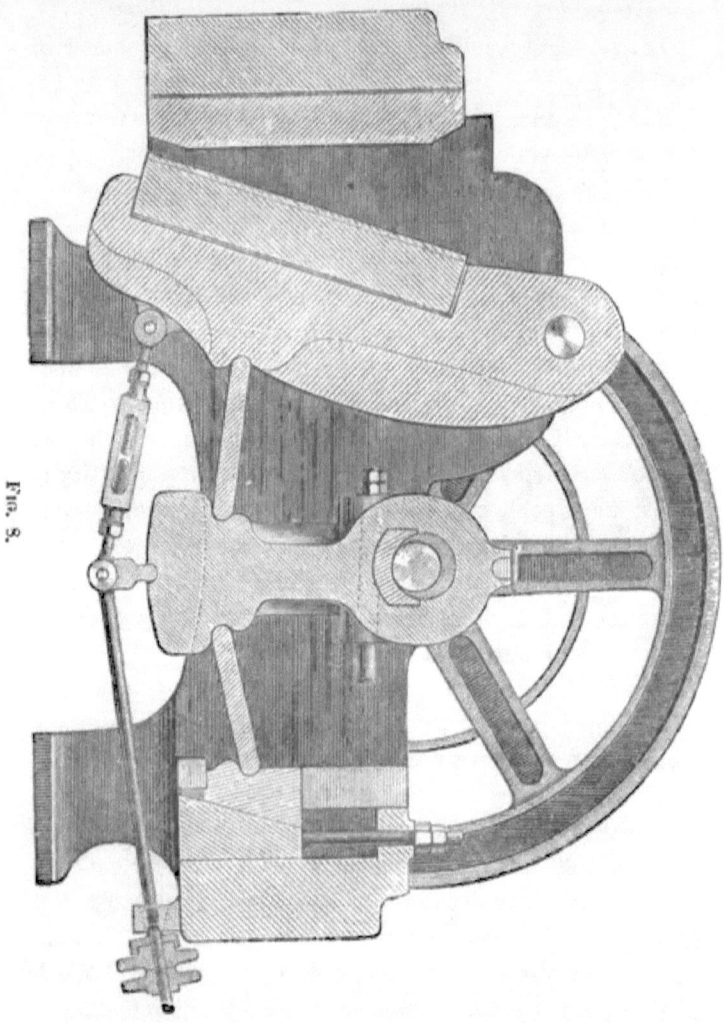

Fig. 8.

or spring attached to the coupling rod X, Fig. 5. The evidence which we have placed before our readers shows that both of the above-named improvers attach much im-

o

portance to the necessity of a more reliable scheme of rebound, and we believe that this apparently insignificant alteration will result in much benefit to those using the Blake or other similarly designed stone-breakers. It is almost unnecessary to remark that the driving shaft of these machines requires to be of great strength and accurate in construction, for unless the eccentric and the head of the connecting or swinging part F (Fig. 5) fit exactly, a great waste of power will arise. The usual method of securing a close and perfect fit is to pour in white metal around the eccentric shaft.

The two machines we have just described differ little from the Marsden-Blake, as they adopt its action and effect the

Fig. 9.

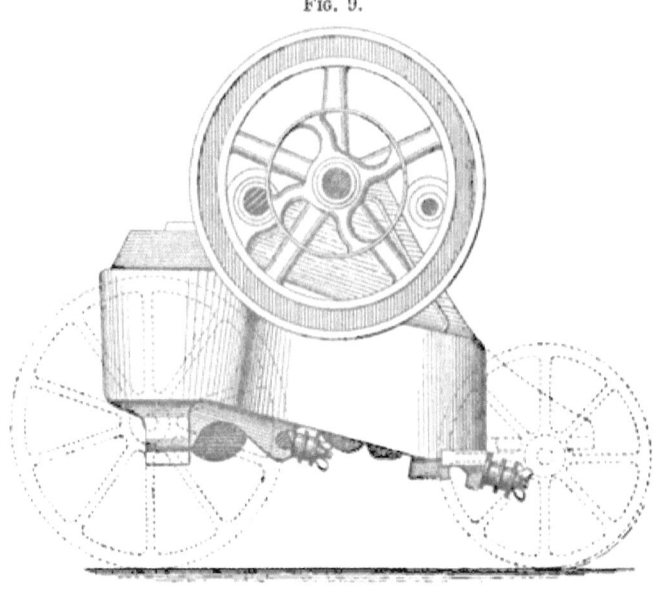

ELEVATION.

improvements on its lines. We shall now proceed to offer some observations and particulars of other stone-breakers differing considerably from these, especially in the mode by which the acting impulse is conveyed to the movable or working jaw.

Figs. 9, 10, 11 represent the "Excelsior" stone-breaker of Messrs. Gray and Co.; Fig. 9 the elevation, Fig. 10 section, and Fig. 11 the plan.

Fig. 10.

It will be seen on reference to the woodcuts that the motion from the eccentric on the shaft is conveyed by a connecting rod to the top of the lever by which is imparted the swinging action again transferred through the toggle to the working jaw. The arrangement for thus distributing the shock must tend to great economy in the working of such a machine. This stone-breaker has obtained a high position in the preparation of *road-metal* for macadamizing roads. For this purpose Hope's cubing jaws are used, and it appears from comparative trials made in Ireland last year this breaker exhibited the most satisfactory results in the production of road metal. In this direction, from a cement-making point of view at least, no special advantage arises, for the more the stone or clinker is fractured the better will be the results. Any quantity of dust can be tolerated by

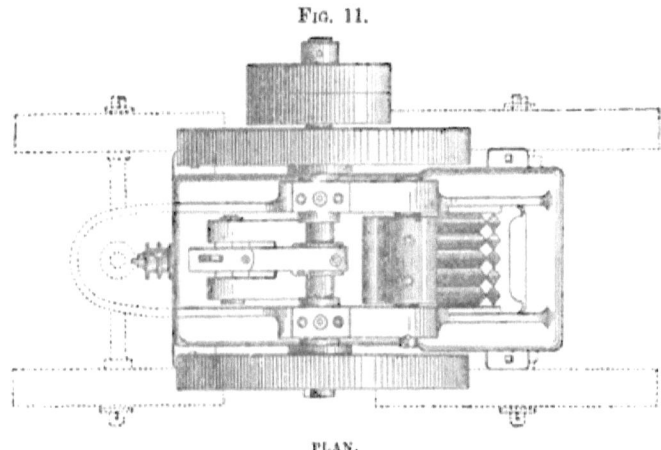

Fig. 11.

PLAN.

the cement maker, for his only object is to reduce his materials as finely as possible, and by the shortest and least expensive means. The "Excelsior" machine is not by its action or mechanical arrangement precluded from producing much finer results than road metal, for it indeed possesses all the facilities in common with the other stone-breakers of reducing the stone to almost any desired size of which this principle is capable.

We will continue our discussion of the stone-breaking machinery by a reference to Archer's combined breaker and pulverizer, made by the Dunston Engine Works Company.

Fig. 12 represents a sectional view of this machine, in which is developed another adaptation of the Blake, that may fairly be termed a compound lever action, dispensing altogether with toggle-plates and similar distributors of the initial force. The centres of the working parts being in one vertical line, reduces the frictional rubbing to a minimum, and as the lever in its return after the performance of the stroke takes the jaw with it, the necessity for the springs or draw-bar action is entirely obviated. Nothing apparently

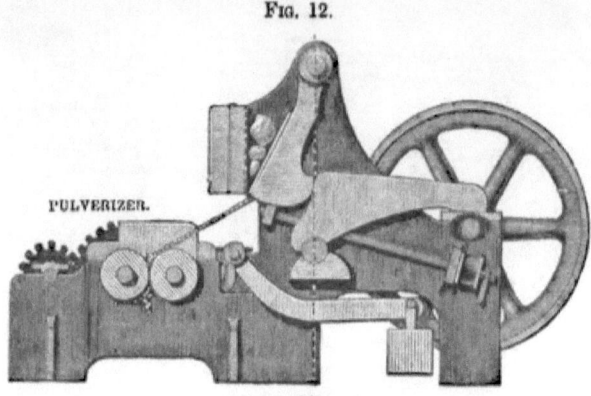

Fig. 12.

PULVERIZER.

STONE-BREAKER.

can be more simple than this breaker, and if the work which it can perform be equal to that done by its more complicated competitors, a step most advantageous to all who require to use stone-breakers will have been made.

The combined pulverizer is not an advantage from our practical point of view of the reducing process, and it would be better, in our opinion, to detach such machines, for they are necessarily driven by distinct and separate gearing.

In this perhaps too lengthened reference to stone-breakers we have purposely avoided alluding to their cost, working performances, or other particulars, as we do not pledge ourselves to either one or the other. The object of this discussion being a desire to familiarize our readers with the distinctive characteristics of the best-known machines. Their several makers can easily be referred to for further information, and this preliminary study through our pages will, we trust, prepare the inquirer for an intelligent selection of what his peculiar requirements demand. It is the duty of cement makers at all events to encourage improvements in these or any other of the machines required by them in their trade, and by so doing give a healthy stimulus to the improver, who can only find the necessary encouragement for progress through such patronage.

There is much dissimilarity in the varieties of the materials with which the cement maker has to contend, and some machines may be found more advantageous than others, according to the quality and character of the minerals operated on. The clinker should be even and uniform in texture, for from whatever materials it may be made, the resulting product should be regularly alike.

The observations which we have made on the various modifications and improvements on the original Blake's machine and its many modifications is partly preliminary to a description of another almost equally useful reducer, known by the name of a "Goodman's Crusher." It is of much more recent introduction than its American prototype, but bids fair to reach an equally satisfactory development.

The Goodman's Crusher or Triturator, which also receives its actuating force from the eccentric action of the driving shaft, is of considerable importance to the cement maker, from its occupying an intermediate position between the Blake's and the millstones, effecting thereby an equal and economical distribution of the work to be performed. Instead, however, of its being transferred by the intervention of other parts or members, it is direct and at once acts on the movable jaws, as shown in Fig. 13.

In our description of this machine we will confine ourselves to the crushers only, as our experience in the use of the triturator for cement-making purposes has not been satisfactory. The combined machine was at first designed for reducing the metalliferous quartz rocks with the view of extracting the ores therefrom, and for this purpose water was

MACHINERY OF REDUCTION. 199

freely used by the arrangement indicated in the woodcut. Probably such a combination of machinery might, however, be advantageously used in the semi-wet process of the chalk

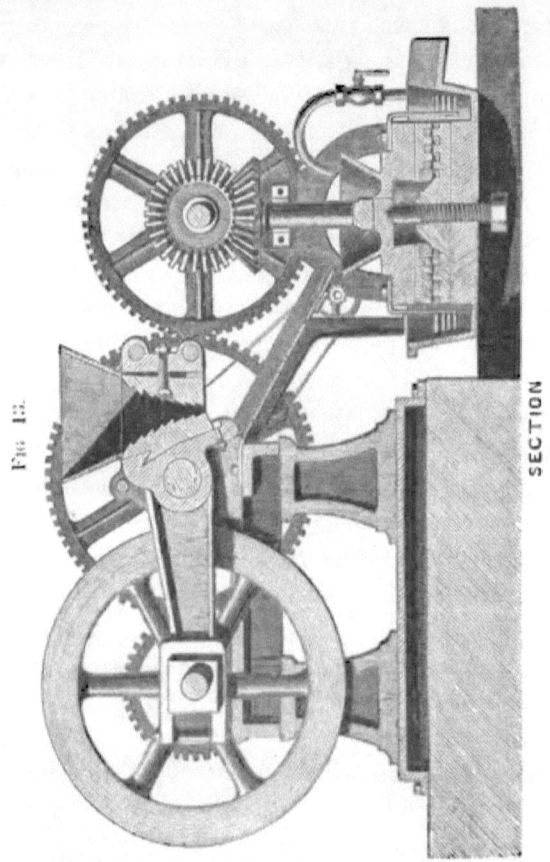

Fig. 13.

districts, known as the Goreham process. The combination involves the necessity of using toothed gearing, and thus much of the advantage from uniform speed and action is lost. There is only a single fly-wheel in the centre of the frame, and the benefit obtained by the double arrangement in the Blake and its various adaptations is thus unattainable.

This machine is simple in character, and has proved itself very useful in receiving the materials from the Blake in pieces of from 3 to 4 cubic inches, or even less. The duty performed should be regarded as an intermediary one, its capacity of work being balanced and its position adjusted so that it may be fed from the Blake, while its produce in turn should go direct to the hoppers of the millstones. This machine can be set so as to reduce the materials to the size of hazel nuts or even smaller, and it would be an advantage to free the output of the "Goodman's" of the dust resulting on its passage to the millstones. A simple vibrating screen or sieve arrangement could readily be devised for this purpose. At all events, and regardless of the advantage which would arise from feeding the millstones with suitable clean materials, the cost and trouble would be more than compensated by the produce from the sieves in the shape of marketable cement on the one hand, and ground raw materials on the other.

The "Goodman's" machine should be driven at about 250 revolutions per minute, and owing to its great simplicity and direct-acting character little attention is required to keep it in true and regular action. It is necessary to be careful that the driving shaft in this as well as in the Blake should be safely protected from the dust, or destructive wear will soon render the eccentric untrue, and the result would of course be increased friction and diminished work. These machines are competent to do equal work with the Blake, only they require a greater amount of preliminary manual preparation of the stones, which cannot be passed through them equal in size to what the Blake machine can deal with economically. It would be putting undue strain on this class of machine to attempt to produce work of an equal character to that performed by the "Blake," and you could not profitably obtain from it in turn the same kind of work as that easily executed by the "Goodman."

The single Goodman's machine is, as shown in section (Fig. 13), mounted on a strong cast-iron frame, having a driving shaft (similar to the Blake), on each end of which is fixed a fly-wheel. The eccentric on the shaft receives the club or hammer, the end of which is produced in the manner shown on sketch. The machine is driven by a pulley fixed on either side of the shaft as in the Blake, and of course the necessity of gearing as in the combined machine is thus entirely obviated. As the shaft or spindle rotates, the hammer is moved backwards and forwards, thus imparting to it a slightly rotating swinging action, the force of which is imparted to the jaw fixed thereto. All the necessary requirements for adjustment of the jaws and their replacement or reparation are equally convenient with the other machines. In those machines first made, diagonally faced jaws were used (that is to say, instead of vertical corrugations they were cut across at an angle of 45°), which arrangement was supposed to take advantage of the peculiar action of the jaw, but in practice they were found unsuitable and abandoned for the vertical ones.

Although we have stated that the action is primarily derived from the eccentric power of the shaft, it differs essentially from the Blake, inasmuch as the force applied is direct and without transfer by lever or toggle-plates. It also differs in being, instead of a parallel thrust like the actuated pendulous jaw, partially rotatory in character, thus effecting a useful grinding action in the arc of the circle which it describes at every revolution of the eccentric shaft. This peculiar action imparts to the machine a pertinacity in dealing with all materials subjected to its treatment, for if once a stone is inserted between the jaws it must eventually pass through, and if unfit to receive at once the necessary impact, the grinding action will ultimately prepare it for its effectual reduction. The Blake machine can only act through fracturing the stone, and until this occurs the jaws must be blocked, as is invariably the case when unsuitable stones are put into the machine.

The balancing or resting of the hammer end or shaft, and the action imparted thereto by the eccentric motion is a weakness in this machine, and adds unnecessarily, in our opinion, to its cost. In fact in nearly all of the machines which we have seen at work this hammer end or shaft, if we may so term it, has been fractured, and when afterwards repaired by bolts and nuts (fastening two parallel plates on each side of the fracture) were found to work more advantageously. It is obvious, therefore, that the too rigid and unnecessarily extended shaft had not sufficient elasticity to withstand the shock of any undue strain on the jaws caused by irregular or ignorant feeding of the machine. It was usually found that an accident of this kind arose from putting too large pieces of stone into the hopper, or the attendants endeavouring by hammering to pass through unsuitable or irregular sized stones.

In Fig. 14 is shown a double-acting Goodman's crusher of a simple kind, and capable of doing a large amount of work if accurately fed and controlled at a proper speed. The action or impact of one side of the moving jaw is downwards and the return stroke upwards, so that only one side of the jaw is under the eccentric impulse at one time. This machine can be adjusted by a wedge-shaped plug or key controlled by set screws at each side of the fixed or side jaws. The parts of this design are few and capable of easy reparation and adjustment, very portable in character, and could be made at a comparatively low cost. Its most desirable situation in a cement factory would be above the hoppers of the horizontal millstones, to feed which its produce could be carried by creepers or screws to the desired points of delivery.

Fig. 14.

DOUBLE-ACTING "GOODMAN'S CRUSHER."

The various improvements in the form and mode of attaching the jaws as now applied to the Blake stone-breakers, admit of similar treatment in these machines or crushers, by which name they are, or at least ought to be, distinguished from the stone-breakers.

A great improvement in the Goodman's crusher is shown by Fig. 15. The machine thus improved has been found to produce much better work at less cost than the original, and its simplicity of construction reduces the weight of the machine itself, so as to render the price less than that usually charged for the present crushers.

An excellent alteration in this amended "Goodman" is fixing the jaw at a lower point, so that instead of the actuating impulse from the eccentric being applied, as in Fig. 13, in the middle, it is placed at the bottom. This difference of arrangement would not have been possible in the original machine, but the guiding or balancing of the jaw admits of this plan, the advantage of which is obvious.

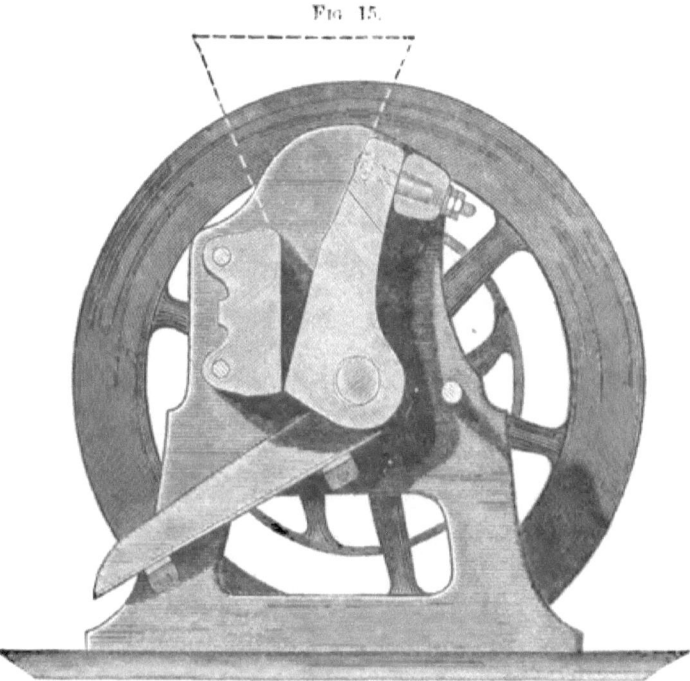

Fig. 15.

There is much difficulty in altering the form of any machine when once it has been introduced, and in the case of the Blake, hedged in by the exclusive patent privileges, improvement was rendered practically impossible. It, however, strikes one at first sight that there is even now room for concentration in the Blake. Machines of this class, required for the treatment of rough and obdurate materials, controlled by labour not at the best of much intelligence, cannot be too simple in character. The eccentric action is of the simplest kind, and has proved, in the extended use of the Blake, well adapted for the exertion of great pressure usefully applied in an indirect manner. The same principle, stripped of the adjuncts of the Blake, is profitably adapted in the Goodman, and still further simplified in that of the "improved Goodman."

In Fig. 16 is delineated another improvement of this machine, for which a patent was granted to Mr. Goodman on the 1st September, 1874. It resembles pretty much the

Fig. 16.

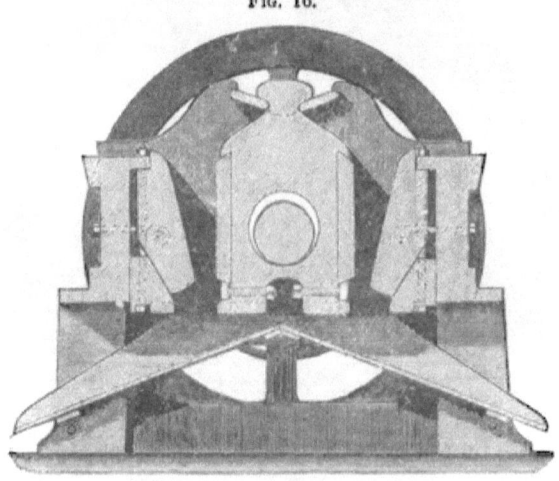

double machine shown in Fig. 14, inasmuch as the movable hammer or jaw strikes alternately an up-and-down blow. The difference consists in placing at the top two toggle-plates, called by Mr. Goodman in his specification *fulcrum levers*, for which also could be substituted friction rollers. As this arrangement appears to us somewhat risky, we will quote so much of the specification as will explain the inventor's ideas on this point. It is as follows:

"The extent of the motion of different parts of the block (movable jaw) is controlled by the fulcrum levers, each of which abuts on one of its sides in a cavity prepared for it in the upper end of the block, and on the other side each lever abuts on recesses formed in the two side cheeks of the frame. The fulcrum levers allow the upper end of the block to rise and fall freely under the influence of the eccentric, but the lateral motion which they permit to the part of the block which is between them is but very slight, but this lateral motion, or motion to and from the fixed crushing faces, increases in passing downwards towards the bottom of the block, and it is greatest at the bottom, where a free escape of the material from the crushing cavity is required, and where the pieces of stone to be acted upon being of small size, a smaller crushing power will suffice. In place of the fulcrum levers friction rollers or other equivalent means may be applied to guide the upper end of the block, but the arrangement shown is preferable."

In the absence of practical experience in the use of this altered machine, the author cannot speak either confidently or with authority, but refers to it here for the purpose of showing that even the Goodman crusher is just entering on its career of change and improvement.

It will be observed that this new machine is the only one of all those we have referred to which strikes the fixed jaw at a different angle, indeed quite reversing the action of the other stone-breakers and crushers. This machine provides the necessary means of adjusting the wearing parts of the jaw or their needful reparation.

We will now proceed to the discussion of the apparatus or machine used in the chalk districts, known as the Wash-mill.

CHAPTER XII.

THE WASH-MILL.

This machine, used only in the wet process, is so simple in character, that any description of it may almost be regarded as needless. At Wakefield, where Aspdin initiated the system of manufacture, and undertook the conversion of materials very different in quality from those subsequently used in the chalk districts, he was obliged to adopt a compound treatment of the limestone and clay—the first being previously calcined and then mixed with the second. By such a process, the wash-mill as now used would have been in a great measure unnecessary, for the finely reduced or slaked lime had already been sufficiently comminuted by the chemical operation of the kiln to ensure its most intimate and perfect amalgamation with the clay. The wash-mill of the chalk cement maker is of the greatest importance in his process of manufacture, and on its accuracy of performance the whole success of the operation depends. The object desired in this treatment of the chalk and clay is the perfect blending of the two materials. In effecting this desirable object the agency of water is brought to bear, and its copious use is indispensable, for the more liquid the mixture, the better the chance of the desired amalgamation. There are several varieties of this machine, from the original harrow with its numerous tines, shown in Fig. 17, to the more modern beam and knife mill, all of which accomplish their task with varying results. In some manufactories it is customary to adopt a double washing operation, and pass the materials through a preparatory mill, from which the liquid is passed on to a second, and from thence conveyed in a finished state to the depositing backs. Again, in other works each material is separately reduced and passed on to

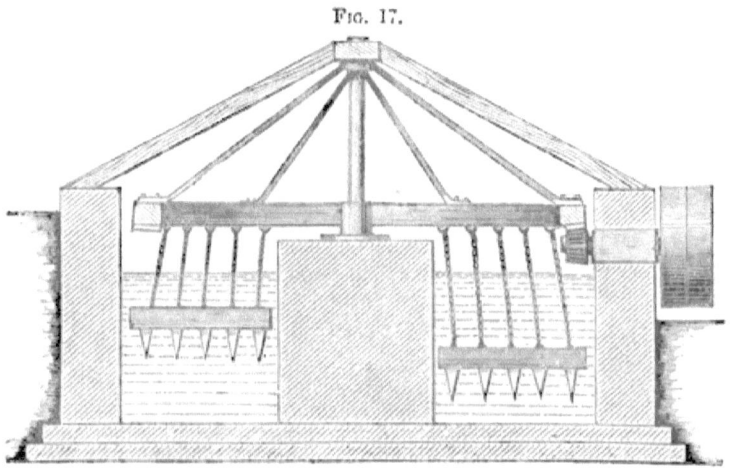

Fig. 17.

a third, where the operation of mixing is performed. There is doubtless much in favour of the double treatment if the accurate blending of the proportions can be secured. The whole process of washing is one which admits of much variety, and the operation is influenced in some measure by the character of the materials to be dealt with.

In Fig. 18 we give an illustration of a wash-mill on the beam principle, the necessary knives or cutters being arranged in the manner indicated. The trough or washing way, as it may be termed, is circular, having a space of 7 feet between the outside and the centre pier, in which is fixed a sole-plate and socket, receiving the toe of the vertical shaft, at the top of which is fixed a bevel wheel driven by a pinion from the driving shaft. The mill should have a substantial foundation of concrete, on which are built the sides and centre pier. The framework of timber must be of full dimensions, so as to be steady enough to resist the vibratory shock of the mill when in motion. The speed at which it is most economical to drive it is about twenty revolutions per minute. The exact speed is, however, a matter best determined by the men working it, whose experience will of course regulate its velocity according to the amount of chalk and clay, as well as the supply of water at command, and which it can economically deal with. The first object to be attained is accuracy of combination of the raw material, and none of the washed liquid should be permitted to pass from

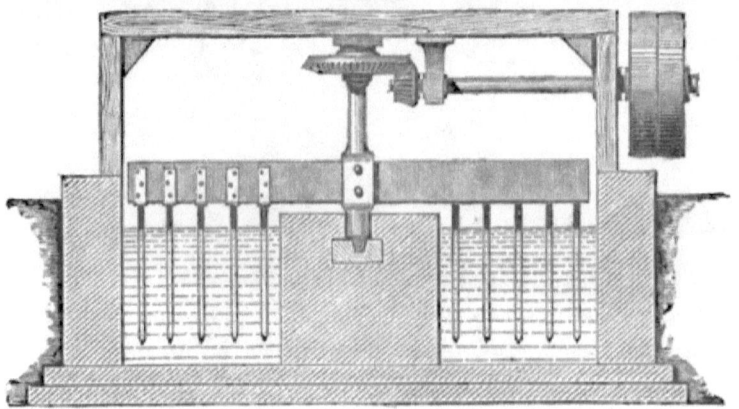

the trough until it has been perfectly amalgamated. The precautions necessary to secure this indispensable result are various, and their detail should be controlled by the circumstances of each particular case. The overflow is perhaps the best method of permitting the liquid to pass from the mill, and at a short distance from the point where this takes place a small pit should be so fixed as to receive the coarser particles of the chalk and impurities of the clay. The wet method involves large back or reservoir space, and in consequence the produce from the mill has to traverse a considerable distance along shoots or conduits, which require constant cleaning, before the liquid reaches its destination.

P

In even the best regulated manufactories the entire avoidance of great waste is impossible, and it is the more unfortunate that this loss is accompanied by an increase of the inaccuracy of result in the backs. The reason is obvious when you examine the nature of the materials and the vehicle of their transmission. The chalk and clay if dealt with first separately would result in a liquid of pretty even consistency, but when dealt with together are, except by a prolonged process, unevenly mixed as well as of varied granular character. The consequence is that while on their passage to the backs, through their sometimes lengthened travel those particles of highest specific gravity are precipitated in the shoots, while the more fluid pass on to the reservoir. Supposing the mixture accurate at its exit from the wash-mill, the chances are very much against its getting to the backs undisturbed and perfect. The more fluid the mixture the less liability to accidents; but this involves not only a large extra cost at the mill, but increases the period in which the products of the backs are required to settle and dry. If the chalk cement makers have an advantage in the favourable character of their materials, they have, in conjunction with them, difficulties that many regard as too trifling, but which, nevertheless, interfere with the profit and comfort of their business. There are three elements of danger in the wet process, one being the careless mixture in the wash-mill; another, the error in passing through the shoots; and the last, the liability to irregular deposit in the backs. These evils are practically unavoidable, and a tendency is now being shown to reduce this wasteful and dangerous process by adopting the Goreham plan of converting the stuff in the following manner. This system has been patented by Mr. W. Goreham, and we cannot do better than refer shortly to it.

The chalk and clay are first passed through a wash-mill

with about one-fourth, or even less, of the water usually applied, and, after as perfect an amalgamation as the circumstances will admit of, are passed on to horizontal millstones (French burrs bedded and jointed in Portland cement). The consistency of the slurry at this stage is of such a character as to permit of its gradual flow, in which state it is received by the millstones, from whence it is passed on direct to the drying plates either by gravitation or otherwise, according to the arrangement of the works.

This process at best only accelerates the time of delivery at the point of desiccation, and thus admits of the further and early conversion of the raw materials, dispensing with the necessity for an extensive system of backs or settling reservoirs, an obvious advantage, and one which must not be overlooked or too slightingly estimated. It appears to us, however, that the wash-mill in this process really after all performs the most important duty, and the millstones only operate in checking the passage of too large particles of the chalk or clay. The time during which the slurry is under the influence of this horizontal mill operation must necessarily be but limited, and therefore liable to produce a mixture somewhat unavoidably faulty in character. In some works, such as that referred to in pages 160 and 161, something like the reverse of this process was attempted, but with materials differing widely in character from chalk and river clay, and resulting, as we have shown, in much loss and annoyance. In other works where the dry process is professedly adopted, it is customary to introduce, after grinding the materials to a fine dry powder, a more than usual quantity of water, so as to bring the particles or atoms into the closest contact. This departure from the true principles of the dry process either indicates an inefficient grinding process or unusually obdurate materials. There is no doubt that a soaking of the materials is advantageous, but should only be tolerated in the absence, or rather the deficiency, of grinding power. Any length of soaking cannot by possibility dissolve a crystalline limestone atom or one of a highly indurated shale, basalt, or granite, and the attempt to supersede the necessity of fine grinding will only result in loss and waste in the succeeding processes.

CHAPTER XIII.

MILLSTONES.

THE cement maker is necessarily largely indebted to the agency of the grinding machinery, and most important assistance is rendered him by the millstones.

The simple action by which their usefulness is made available may be likened to the familiar example of the scissors, the cutting operation of which is due to the two blades acting against each other. So in the case of the millstones, the fixed or stationary bed-stones being dressed into furrows, are crossed during their circular traverse by those of a corresponding character in the runner or upper stone. The diagram of the "dress" (technical term for arrangement of the furrows and their repair and renewal) is not arbitrary, but rests in a great measure with the operative millstone dresser, guided by his practical experience and knowledge of the materials which he has to reduce.

The stones are of various kinds of distinct mineralogical character, and are obtainable from different geological formations, but those most suitable for cement grinding are the well-known "French burrs"—commonly and indeed, we believe, solely used by the cement makers in England—and the celebrated lava deposits at Andernach, on the Rhine. The origin of the term "burr," now perhaps in milling phraseology exclusively applied to the stones brought from France, had another and more ancient origin. The *burr* was the rough dress of the stone, and at the time that millstones were exclusively obtained from the *millstone grit* and other kindred formations, those stones which maintained their roughness, or "burr," longest were the favourite stones, and most generally used. In those days, however, it was only corn that was ground, and therefore the duty required of them was comparatively light. The work of cement grinding is a very different business, and the selection of the proper stone for that purpose is a matter of much importance. The building up of cement stones is also a work of considerable accuracy, and those makers who give this branch of their business special attention, are well rewarded by the good reputation they obtain. Such makers personally select the stones at the quarry in France, where they are obtained, and thus avoid the risk of sending inferior stones to England, the freight and land transit of which add so much to their cost.

These stones are obtained from the tertiary deposits of France (Seine-et-Marne), and are of various shades of colour, from a bluish white to yellow, and reddish. The first is the most suitable for cement stones. They are almost pure silica, with a slight amount of calcareous matter, and are honeycombed, in appearance being slightly porous; flinty in their physiographic aspect, and much resembling a conglomerate of flint splinters, with the cavities and spongy deposits common to that mineral.

The experienced millstone dresser acquires a knowledge of the best stones for the work, but even the most carefully selected stones exhibit sometimes internal flaws and fractures which materially affect their value. It is safer therefore to put implicit faith in the stone builder, whose best interests are identified with the durability of his workmanship; and when a faulty stone becomes prematurely useless, the maker readily replaces it for the sake of his reputation.

The expense of this department of cement making is considerable, and entails an amount of care and watchfulness which if neglected adds materially to the cost of manufacture. It is, in fact, a never-ending expense, not only in the wear, but the continually recurring dressing done by experts, who are of necessity costly in their maintenance.

Stones are built up in two ways, viz. the burrs are laid on their beds in one way and on their edge in the other. Fig. 19 shows how they are built. It is unwise for cement work to have any but those built on edge, and especial care

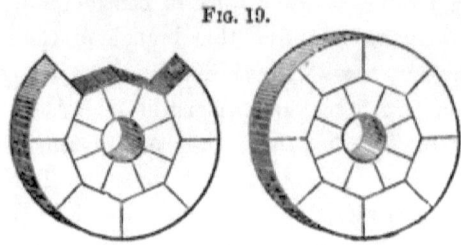

Fig. 19.

should be taken to have eye burrs as large as possible, as it is there where the greatest amount of wear and tear takes place. These burrs should be of extreme depth and have finely dressed joints. In fact every joint should be carefully made, not only at the surface of the stones, but through their whole depth. The reason for such accuracy is obvious, for as the stone wears the joints widen and involve the necessity of filling them with lead or other metallic substance. It is customary for some manufacturers to inspect the jointing of the burrs before the final building up of the stone. This, however, may be regarded as superfluous where the stone maker himself runs so much risk. It is his interest to secure sound work by his personal supervision; stones are, however, generally built by task work, and the keenest vigilance is not always competent to detect imperfect and unduly hurried work.

Backing the stone to increase its weight, and strengthen it by hoops and bands, is the finishing operation. The concrete backing is composed of chips of the *burr* and Portland cement, and it is better that the stone should be finished for at least a month before being used. The stone maker supplies the stone in a perfectly level state, and the further preparation of its surface is left to the cement maker.

Of late years much alteration in the dress has been occasioned by the introduction of intermediate machines, such as stone-breakers, rollers, crackers, and triturators, and it will depend pretty much on the condition in which these various machines deliver their produce into the millstone hoppers what form of diagram the stones should assume. Great attention is necessarily directed to this stage, for on its accuracy the economical action of the stones depend. If the clinker is reduced to the size of horse-beans, a small amount of swallow (the hollowed space round the eye of the runner) will suffice, and therefore a more extended surface is left for the skirt (the furrowed remainder of the stone extending from the edge of the swallow to the periphery of the stones), where really the most profitable part of the whole finishing operation is performed.

It would be impossible to prescribe any rule or furnish arbitrary diagrams for regulating the furrow dress, where the conditions by which it must be regulated are so varied. It is enough to say that practice will determine the best form of diagram, which should be so regulated as to deliver the finished material from the stone as fast as it is produced. The speed at which the stones revolve is a necessary element to be considered in such a calculation, for if the stone delivers too freely the coarse will pass out with the fine, and so render the result unsatisfactory. On the other hand, it should be borne in mind that finished dust if retained in the stones tends to retard trituration, or reduction of the coarser particles; its fineness being interposed between the rough pieces keeping them from coming into economical contact with each other, as well as preventing their being acted on by the grinding surfaces of the stones themselves. So much has this been regarded at all times and for all purposes as a defect in the present scheme of grinding, that numerous plans and devices have been used to separate, by blowing or otherwise, the finished from the unfinished material. All and every such application has, however, resulted in more or less disappointment; at all events, the writer is not aware, in the production of cement at least, that any profitable arrrangement for such a purpose has been carried out with success.

It must be apparent that after every condition of accuracy—the original selection of the burrs, their careful building together, followed by the approved dress—has been performed, that the most important operation of all, that of putting them in their true relative position, has yet to be considered, as shown by Fig. 20, p. 220.

As a preliminary to our description, we will assume that the spindle C has been accurately fixed, so that the lever acting on its shoe is capable of instantaneous action so as to secure the accurate adjustment of the stones.

The bed-stone, F, is first fixed truly level on the place prepared for it on the frame or hurst, the neck arrangement for permitting the free rotation of the spindle, and carefully packed to prevent the dust from passing through on the gearing below, as well as to prevent damage to the spindle itself. The greatest care should be bestowed in putting the bed-stone exactly level, and seeing that its upper or working surface is at right angles to the spindle. The hanging of the runner, E, as it is technically termed, is an operation of extreme delicacy, and it should be performed so that the runner shall rotate exactly parallel to the bed-stone; not in an intermittent or wobbling manner, but uniform and exact. The thimble on the top of the spindle is usually made of steel, and reduced as fine as possible, so as to be consistent with a true reception of the indentation in the cross-bar, H, of the runner. Many an accident is due to the careless performance of this simple operation, and much loss is thus occasioned by the unsteady rotation of the running stone.

Description of the Millstones, Framing, and Gearing.

It may perhaps be regarded as superfluous to minutely describe a machine whose parts are of so well-known a character, and one that in various modifications has existed for man's necessities since the advent of civilization itself. As this work is intended, however, to deal with the whole subject of cement making in its most comprehensive form, it will not, we hope, be considered as a waste of space to offer particulars of one of the most important mechanical aids in that manufacture; for in both stages of the process — first, the perfect conversion of the raw materials by the dry process, and the succeeding and final operation of the conversion of the products into a fit condition for constructive and other purposes—it plays an important part.

In large manufactories, especially those of old construction, the system of continuous shafting has been generally adopted, and at the proper interval the mitre wheels are placed from which the motion is imparted to the pinion fixed on the stone spindle. This system of arrangement of the grinding machinery involves much outlay in solid and substantial foundations and costly framing to withstand the attendant vibration when in motion. In very large mills this is sometimes a serious difficulty, and in more than one instance known to the writer, a large part of the machinery in an extensive mill had to be disused owing to the danger and inconvenience consequent on the vibration caused by the whole of it being at work at the same time. When the transferred power is direct by gearing from the motive or engine power, this objection exists in its most aggravated and dangerous form, and should only be tolerated when its necessity continues to exist through old arrangements, or the difficulty of substituting the more portable system which we are about to describe.

In either arrangement of the driving power, when conveyed from the engine, it is of course necessary to have main shafting to transfer the power by driving belts to the various grinding machines: in the one case, by having the shafting fixed underneath the floor of the building; in the other, for portable machines, at a convenient height in the mill itself. These arrangements are, however, subject to such modifications as can only be controlled or regulated by the requirements and circumstances of each particular case.

The frames, whether fixed or portable, are identical in their purpose, and the following descriptions of the stones and their gearing are equally applicable to both systems.

For convenience of illustration, we will refer to the woodcut of a portable hurst or frame, Fig. 20.

The horizontal shaft fixed in pedestals on the base of the framing receives its actuating force by a belt on the fast pulley, and can be transferred or thrown out by shifting the belt on to the loose pulley adjoining. The motion to the spindle, C, is conveyed by the mitre wheel shown in section, the bottom of which rests in a step-piece placed on a bridge through which runs the horizontal shaft, and connects the lever arrangement that regulates the rising and lowering of the stones, E E. This arrangement is of the utmost importance, as through its agency the regulation of the grinding and its quality are assured. F is the bed-stone, B the casing, A the hopper in which the materials to be ground are placed.

It is necessary, as must be obvious, that the spindle during its rotation in the bed-stone, F F, should be so adjusted as to prevent the passage through of any dust, however fine. The arrangement for the effectual accomplishment of this desideratum is thus:—In the centre of the bed-stone is placed a cast-iron bush or box which provides for the introduction of packing composed of various substances, that effectually prevents any of the dust from passing through, and at the same time secures a minimum amount of friction on the

Fig. 20.

spindle. On the remaining portion of the spindle is accurately fixed a cast-iron cross-head, on which is hung the cross-bar of the runner.

On the top of the cross-head, where the two pins shown on Fig. 20 are placed, the damsel, or apparatus for beating and regulating the supply to the stone, is fixed, and plays between the shoe or shoot which is placed under the hopper, A. The adjustment of this simple part of the machine is secured by a piece of steel which acts as a spring, and its pressure may be increased or diminished according to the capacity of the millstone to receive the contents of the hopper, or, in other words, to feed them with regularity and not overload them, which leads to congestion, stoppage, and, when carelessly arranged, breakage. The nice point to be considered here is that the stones shall receive just as much, and no more, than they are competent to reduce. When freshly dressed, the amount of work is greater than when worn and smooth, for the furrows are then more or less worn down, when their grinding capacity is sensibly reduced.

From the above description it will be seen that the runner rests on the centre of the spindle at its point of contact with the cross-head fixed over the throat or swallow of the stone, so that practically the brass shoe or socket receiving the bottom of the spindle supported by the lever sustains the whole of its weight. This, then, should be regarded as that part of the machine requiring the greatest nicety of adjustment and attention, as by its means the accurate rotation of the runner or upper stone is secured.

All the salient points in the economy of this machinery must of course be familiar to the experienced mechanic and millwright, but it is also important that those having the control of its working should understand its leading features and characteristics, so that when accidents or disorder arise they may be competent, if not to repair, at least to direct the necessary rectification. At all events, a sensible knowledge of even the minutest detail will not lessen, but, on the contrary, increase the efficiency of a director or manager of cement works. It is the ignorance of the employer or manager that leads to the dogmatic and sometimes overbearing conduct of the skilled workman, who too frequently acquires his knowledge from purely practical sources, without any theoretical or scientific basis whatever.

Before leaving this part of our subject, we would shortly refer to the desirability, in small, or indeed in any, manufactories of adopting the portable or distributive system of arrangement of the machinery. When large and powerful engines are employed, the distribution of the power to the various points involves a great loss in friction, and also leads to much difficulty when a break-down of the engine, or repairs or replacement of boiler, causes a stoppage of the entire machinery.

Cement making is peculiarly well adapted for the adoption of the divisional system, from the unavoidable disjointed character of its operations, and such an arrangement may be applied with advantage, more especially when the necessities of the dry mode of manufacture involve the duplication of like machines, so that if 50 H.P. was required its division into two 25 H.P. would afford a more economical arrangement, besides the advantage of being able to work one set of machines only when necessary.

It might be as well to allude here to the tools used for dressing the millstones. They are simple in character, made of the best cast steel, and consist of two kinds, bills or picks, and pritchells or chisels. They vary in weight according to the character of the work in which they are to be used. The bills are fixed in the thrift or handle, and the pritchells are worked with a hammer.

The exact velocity at which the stones should revolve is a matter which depends in a great measure on their size; the speed beng regulated by the diameter of the stones: the smaller ones requiring to be driven quicker than the larger ones, about 25 feet per second being regarded among practical men as a fair average, and economical as well as safe velocity.

There have been many attempts made to contrive a machine capable of grinding corn and minerals as fine as the produce from the horizontal millstones of which we have endeavoured to give an intelligible description: conical mills of a variety of designs, corrugated steel discs running horizontally, and numerous other devices, resulting generally in much dissatisfaction to those induced to speculate on their untried merits. In Portland cement making the necessity of extreme fineness practically prescribes the use of the French burr stones, which by continual dressing are competent to produce the desired and, indeed, almost if not quite, indispensable results. The unavoidable wear and tear and attendant cost of the maintenance through manual labour of a rough burr or dress in the stones renders this department of the manufacture one of considerable cost. When sieves are used and tolerated in a manufactory, it is not so necessary to aim at extreme fineness of powder from the millstones, but we do not regard this arrangement as at all desirable, and prefer to accomplish the best duty from the grinding machine of which it is capable.

Hitherto our ideas have been restricted to the action of the stones doing duty in the old and original horizontal position, and generally, as a rule, expected and indeed believed the best economy was derived from being able to load the running stone with a considerable additional backing or weight. In the accompanying woodcut, Fig. 21, is what is called a "scientific grinding mill" by its makers, Messrs. A. W. Straub and Co., of Philadelphia, United States of America, capable of doing increased duty, and which, in an experience of some years, has superseded the horizontal millstones for all those purposes to which the burr stone can be applied. There is no pretension in the claims of the makers to supersede the use of burrs, but merely in the new mode of their application to drive them vertically instead of horizontally.

The description of the mill is as follows:

The *Bed-plate*, supporting both journal and bed-stone, is cast in one piece, like that of a steam engine, to keep all the parts in perfect line. It has the stone-case in the middle with

Fig. 21.

large dirt spaces at each end, allowing all dirt escaping from the stone-case to fall on the floor without passing into the journal boxes. The space around the stones is $2\frac{1}{2}$ inches, and the discharge from the bottom, so as to prevent clogging.

The *Bed-stone* is fast in an iron ring, bolted against the end of the stone-case, forming a dust-tight joint; it is framed to the runner with three set screws.

The *Journal Boxes* are 7 inches long, have caps to take up all wear which may occur. They are lined with Babbit metal to make them durable.

The *Spindle* has the turning stone cast fast to it with zinc;

the screw conveyer, eccentric, and pulley are also fast to the spindle.

The *Temper Screw*, at one end of the bed-plate, rests against the spindle.

The *Pulley* has a hub upon one side, the journal box passes half-way through on the other side, to remove the strain of the belt from the spindle; this allows the belt to approach from any angle, and be removed without unscrewing.

The *Legs* raise the mill from the floor, allowing it to discharge from the bottom, and saving the expense of a foundation.

The *Hood* rests upon the bed-plate, with a heavy twine packing laid between them, a large square feed-trunk at its end extending downwards.

The *Feed Shoe* is damzelled in front by the eccentric on the spindle. The *Hopper* has a valve in the bottom to control its discharge.

These mills are made with burr stones from 12 to 30 inches in diameter, and are driven at speeds of from 200 to 800 revolutions per minute, the power required being from 6 to 30 horse-power. Of course power and velocities are regulated by the kind and quality of work to be done.

The price of the machine is not a high one, and if it can do the work claimed for it, there is a possibility of adding another auxiliary to the cement maker's appliances. In very extensive works, probably a light machine of this character would be found unsuitable where a large number of stones was required, and where it would be necessary to arrange a complete system of shafting and its attendant expensive gearing. In small and especially experimental works, a grinding machine of this character would doubtless prove a great advantage. The small space which it occupies and its portable character are great recommendations.

The mode adopted for dressing the stones when required is the following:

Remove the hood, journal caps, pulley, and two bolts which secure the bed-stone to the end of the stone case, raise both stones above the stone case (with a crane and screw), place two boards across the bed-plate, lay the stones on these boards face up.

With a sharp mill-pick dress the face of the stones in fine lines, crack them rough and deep. The furrows should be cracked and kept half an inch deep at the eye, and shallow up to one thirty-second at the skirts to prevent the stones throwing grit. The sharp feather edge should be shouldered down one thirty-second of an inch. In this mill an extra pair of stones may be at hand so as to put them into their frame, and thus save time while the blunt or worn pair are being dressed or sharpened.

It is thus necessary to describe this grinding machine and the mode of dressing it, for the high speed at which the stones rotate involves a special arrangement of the furrows as well as the form and shape of the eye. In such a machine it would be necessary to feed with small pieces not larger than peas, and we imagine (from our imperfect and slight knowledge of this novel grinder, we are unable to speak authoritatively) when so supplied much effective work might be realized.

In some of the conical mills to which we have alluded burr stones were in several of the modifications used, but the difficulty attending their dressing and the smallness of the grinding surfaces prevented those mills from obtaining any permanent or useful position.

In America and other profitable gold-producing countries it is a matter of primary importance to be able to extract economically the precious metal from its matrix. In the localities where the auriferous rocks prevail, labour is usually costly, and therefore the greater necessity for machinery competent to assist in the extraction of the metal from its native rock. Some years ago a machine was patented by a Boston (United States of America) Company, of the following description.

It consisted of a massive revolving table, 3 feet 6 inches in diameter, travelling at a speed of upwards of a thousand revolutions per minute. It was furnished near its outward circumference with cutting or splintering blocks made of specially hardened metal. On the entrance of the material to be reduced it is driven from one side to the other, and receives a series of frequently recurring blows from the revolving blocks, and when reduced sufficiently passes out at apertures in the side of the tub or casing. This machine is costly in character, and was represented as competent to reduce gold quartz to the size of coarse gravel at the rate of 200 tons in twenty-four hours. The size or quality of the products from this machine was capable of regulation by the form and extent of the perforations at the circumference of the casing. The machine was a complicated one, and if metal could have been procured durable enough to withstand the shock and concussion of a so violently propelled mass of metal, its success might have been secured. The result, however, of all this ingenuity was that the machine, while reducing minerals, could not prevent its own degradation, which became so great as practically to render it useless.

The "Carr's Disintegrator," in use for reducing minerals of various kinds, and even grain, obtains its value from the violence of the action produced by bars revolving in opposite directions, against and between which the materials under treatment are forced, coming also into contact with each other, the high speed of this machine, as well as the American one, producing draught or ventilation, thus expelling the finely comminuted particles as they become small enough to pass out. This class of machine is not, however, reliable enough for cement-making purposes, and the high speed at which they must of necessity be driven incurs liability to accidents accompanied by excessive wear and tear.

What the cement maker requires is a machine competent to perform in a united shape the work of a "breaker," "crusher," and "grinder," and the author feels assured that a machine of this kind will be forthcoming through some such agency as Sholl's direct-acting pneumatic stamper, which appears to possess the necessary arrangements for a steady reliable blow either by impact or percussion. It is low-speed machines that are wanted, and not those which are violently driven and uncontrollable.

CHAPTER XIV.

BACKS OR RESERVOIRS.

This arrangement for receiving the produce of the wash-mills is peculiar to the wet system, and their proper arrangement has a marked influence on the economy of the chalk and clay process. In the early manufactories they were generally placed on a level with the surface on which the works stood, and their contents when in a fit condition wheeled on to the drying ovens or plates. In the most economically arranged factory the wash-mill is so placed that it shall be convenient for the reception of the raw materials, and at the same time at such an elevation as will permit of the flow of the washed products by gravitation to the backs, a main shoot or leading conduit traversing the whole length or circuit of the backs, all of which are filled from it by the agency of subsidiary channels of minor capacity; the drying ovens, when most conveniently placed, being near the kilns, and, when possible, adjoining them, so that the produce from the plates when dry can be in a direct manner at once hoisted into the kiln. The backs under such an arrangement must necessarily command the means of easy drainage, not only for the overflow of water in the ordinary process of decantation, but in addition should be so constructed as to permit of this being done readily in any and every direction. The strength and form of walls will in a great measure depend on the depth and capacity of the backs as well as their position. Concrete walls are the most economical, but as a well and substantially built wall of this kind is less porous, or rather possesses less capacity of absorption than bricks, convenient spaces at frequent intervals should be provided where drainage of the water could be encouraged. The foundations or bottoms of the backs should be rendered as porous as possible, for if it is completely water-tight and impervious to moisture, the water of the slurry can only escape by the pressure or squeezing of the moisture, a slow and dilatory process even under the most favourable circumstances. The size of the backs can only be determined by the extent of the manufactory, and it is advisable, unless where circumstances render it impossible, to make them of a medium size, for it must be remembered that their filling is necessarily a slow and tedious process, and even under the most favourable condition of the weather involves a long interval before the contents are ready for the drying plates. As by the washing process there is no further preparation of the washed slurry—such as is the case in the dry process where the powdered mixture is rendered plastic enough to admit of being moulded into bricks—the best results are those which admit of the driest material being placed on the plates, and thus avoiding the necessity of evaporating a large amount of moisture. In well-regulated and economically arranged works the back-room should be so balanced as to secure the most favourable condition of the slurry at all times during every season of the year. You may commit the error of having the produce from the backs too thin, but you never can be blamed for its being too dry. The weak point of the wet system of manufacture is undoubtedly in this department, and when it is unduly pushed the profits must diminish, and indeed total disorganization ensue. Washing the materials is an operation which under no circumstances admits of careless treatment; burning the kilns cannot be prejudicially hurried; grinding the cement must be accurately performed; but the conduct and controlment of the backs are subject to the most eccentric treatment, and when completed the loss incurred is beyond recall and retrenchment. The existence of the backs furnishes the best argument against the wet system, and the fact that the London makers are turning their attention to their supercession by other means less costly, goes far to prove that they are wrong in principle and extravagant in maintenance. As chalk contains a large percentage of broken and almost invisible fragments of shells, on which the best influence of the washing apparatus has but little influence, it is required that the means for preventing its passage into the backs should be as perfect as possible. Perhaps the best plan for trapping, as it were, this useless and indeed dangerous material is to provide a small catch receptacle at each back, into which it would gravitate and be intercepted. The sand, which in some deposits of chalk is very considerable as well as that from the clay, must also be caught through the same or similar agency. In some *chalk and clay* cement works conducted under the wet or washing system, as much as ten per cent. of this waste arises. In the dry process it would be inappreciable, or pushed to such a degree of fineness by the more perfect grinding process, as to reduce its injurious influence to a minimum.

In connection with the subject of backs, and one intimately identified with their economy and usefulness, is the important question of the density of the washed material. It is of so much consequence that the calculation of the size and form of the backs should have direct reference to their capacity or power of rendering the contents as compact as possible. The deeper a back is the more efficient will its power of compression of the materials be, and it is preferable where space and time are controllable to insist that the backs should be deep rather than shallow. An examination of a piece of dried material as it passes from the drying plates to the kiln will exhibit an amount of porosity almost incredible. These interstitial voids are the spaces or points occupied by water, which on the application of the drying heat becomes converted into steam, causing the cavities of an irregular character to interpose between the particles and prevent their true and economical action in the kiln. This is the reason why air-dried slurry is better than that which is heat-dried, because by the former process the particles are more intimately connected, and the dried mass much denser in character. When we discuss the brick pressing and forming machines, this advantage will be more apparent, and we will endeavour to show that the more compact and dense the raw material can be passed into the kiln, the greater will be the advantage derivable from the heat to which it is subjected.

Much has been said about testing the contents of the backs, and specially designed apparatus have been devised for that purpose. This practice, however, even when conducted under the most favourable circumstances, can only be deceptive in character. When too much reliance is placed on this mode of rectification, the more important points of challenge are neglected. The most perfect and reliable method of admixture of the chalk and clay would be by dealing with them in separate wash-mills, and maintaining a uniform quality and density of liquid, so that both fluids could be volumetrically gauged at their entrance into the final mixing mill.

For so far our observations have been made with reference to backs and their being charged or filled by the simple and inexpensive gravitation method from the wash-mill. The economy and arrangement of some manufactories necessitates the elevation of the products from the wash-mill by force-pumps to a level high enough to permit of their contents being wheeled on to plates heated by the waste gases from the kilns. In the first-designed arrangement of this kind considerable expense was incurred in the erection of brick arches to receive the drying plates, but the arches themselves served as useful receptacles for the storage of coke and other purposes. By Johnson's process the washed material is still elevated, but into high cylindrical reservoirs of moderate capacity, from which it flows, after sufficient decantation of the water, on to the drying plates or channel heated by the waste gases from the kiln.

In the dry process no great amount of storage for the powdered raw materials is required, and in well-arranged works two, or at most three, hours' supply only is needed, unless where it is necessary to soak the raw materials; in such cases twenty-four hours' stock will be ample. In the description of the process adopted in the manufacture of Portland cement from the blue lias materials at page 171 we say that the powder is passed down from the dust-room to double screws, by which the requisite amount of moisture is evenly and regularly applied on its passage to the pug-mill of the brick-forming machine. The blue lias materials are, comparatively speaking, soft and smooth in texture, so that this process is enough, but in other more crystalline minerals such treatment might not prove sufficient. A remedy will, however, be found for a partial if not entire mitigation of this difficulty by using the machine for pressing bricks described at page 266, and we hope to see it or some modification of its principle superseding the ordinary types of brick-forming machines.

CHAPTER XV.

DRYING OVENS.

THE continually changing character and improvement in the manufacture of cement must eventually result in the abandonment of this arrangement for drying purposes. As, however, small and experimental works may be so circumstanced as to prevent the possibility of adopting the new plan, we consider it necessary to allude shortly to the drying plates as the means of obtaining the proper heat for the desiccation of the raw materials.

The conditions which regulate the erection of these buildings are in a great measure dependent on the locality of the works, and their relationship to the fuel supply. When a good and unfailing quantity of cheap coke (of the required kind) can be commanded, the drying plates need not be of a substantial or costly character, but may consist simply of a series of covered flues heated by any waste or exhaust steam, or the cheapest coal or breeze capable of imparting the required heat. It is advisable that they should be, in any case, built four or five feet above the level of the ground, upon a good bed of concrete, so as to prevent the absorption of moisture, which would prejudicially interfere with the action of the flues and their profitable working. The first cost of such drying plates is moderate, and, when carefully regulated, easily kept in working order. Their extent will entirely depend on the capacity of the works; but, at least, they should be competent to receive all the slurry from the backs, and be able to supply the daily requirements of the kilns.

Another condition is when provision has to be made for making all the coke required for burning the cement. In such a case there must be substantially built ovens capable of coking as much coal, in the twenty-four hours, as will supply the kilns with the proper amount of coke. Twenty-four hours' coke, if made from a suitable coal, is strong enough for cement-burning purposes. Although the primary object is the attainment of coke, care must be taken that the ovens are performing this important duty so that the heat evolved during the operation shall be economically utilized and applied to drying the slurry. The ovens should be constructed of a size to coke from 20 to 30 cwt. of coal in the twenty-four hours, and be so connected with a shaft that, by dampers or similar arrangements, the proper draught can be maintained at all stages of the process. There are, of course, great varieties of coal, many of them unsuited for coking, except under most expensive conditions, and it will be advisable, therefore, to obtain a coal bituminous enough to produce coke. The coal should be decided on before preparing the plans of the ovens, for on its quality will much depend their size and form. Unless in London, or the coast, or other similarly favoured locality, where Newcastle coal can be cheaply obtained, whose conversion into coke is easy and simple, no great difficulty surrounds this operation. There are very few coal-fields indeed in this country from which the obtainment of a coking coal is impossible, but the qualities of even the best are so variable that every care must be taken that the form and capacity of the ovens should be such as meet the peculiarities of their varied chemical and mineralogical character. An oven arranged for the coking of a first-class quality of Durham, Wigan, or other equally good Lancashire coals, would fail to produce the same economical results from a Derbyshire, Leicestershire, or Warwickshire coal. There is one important consideration which should not be lost sight of in the selection of a coal from which coke is to be produced for burning Portland cement, and that is its freedom from ashes. If this is neglected, and a spongy, ashy coke produced, it will in a most prejudicial way influence the quality of the cement. The more free, therefore, the coal is from objectionable mineral ingredients, the better will be the coke.

The primary object of coking coal is to obtain a concentrated fuel of intense heating power, and, for foundry and other smelting purposes, it is essential that the process should be sufficiently prolonged to purge the coal of its impurities and realize the densest and brightest crystalline coke. To produce coke of this character, specially constructed ovens are required, and the heat evolved in the process is thrown off in waste. The utmost value of heat resulting from coking is, however, obtained with the least waste of the coal from the gradual character of the process performed.

The waste, or perhaps, more properly speaking, the discharged gases from the coal, is on the average, under favourable conditions, from 40 to 50 per cent. of its weight, and for cement-making purposes we may take the former amount, as the process is not so prolonged in that case. However, the heat evolved cannot be regarded as lost, for it is utilized in heating the drying plates which receive the slurry from the backs. In the ordinary coke ovens used in cement works it is hardly possible, in the restricted time allotted for coke production, to fully purify the product from the coal, and it is therefore best to avoid the use of those qualities with a large percentage of iron pyrites. The injurious influence of sulphur in the composition of a Portland cement is well known, and in some of the raw materials it exists in sufficient quantities without adding to the danger by permitting an excess of it in the fuel of the kiln. Some of the strongest and hottest of the Derbyshire coking coals are so impregnated with iron pyrites that the author has had to direct their disuse in some cases, although the coke produced from them was hard and free from ashes. From what we have said it is apparent that the cheapest coal is not the most economical for conversion into coke for cement burning; neither, in the end, will it be found the most useful for warming the plates for drying purposes.

In those factories that are dependent on the kiln fuel from the ovens, it will be advisable to arrange their position most convenient for the receipt of the coal and close to the kilns. It will generally be found that their most economical site is between the backs and the kilns, and their extent should be regulated by the necessities of the works. The best form of oven is about 10 feet long, 5 feet wide, and 4 feet high, having the arch of a well-balanced segment of a circle, neither too high nor too flat. When built in groups of six, they are convenient and economical. Care should be taken that the foundations are well protected from ground moisture, and either a thick bed of concrete or hollow arches should be constructed underneath them. They should be lined with ordinary fire-bricks at the sides and ends, but the arch must be turned in the best fire-bricks. The arch should be perforated with $2\frac{1}{2}$-inch holes to allow some of the heat to escape into small flues on the top of the arch and through its spandrils. On these flues are placed the best fire-tiles, and for at least 10 feet beyond the ovens, as it is here that the most intense heat occurs on its first entering the flues. The remainder of the flues may be covered with cast-iron tiles half an inch thick. The length of the plates are of course much dependent on the duty required of them, but generally speaking they will be found most economical and useful when about 50 or 60 feet, excluding the ovens. The best arrangement for the flues is to convey the waste heat from the ovens into a main cross-flue parallel to their ends, connecting it with the various longitudinal flues, which again should discharge into another large cross-flue running into the chimney. At the junction of this main flue with the chimney, the damper by which the draught is regulated is fixed. The process of coking fuel requires the exercise of a certain amount of intelligence, so that when the oven is first lighted the draught should be carefully regulated, and no more air admitted than is wanted to ensure the proper ignition of the mass. So soon as the body of coal is fairly alight, the access of air should be completely excluded, and the damper at the chimney regulated to the required point. In the flues it is advisable to place, at certain intervals, *soot traps* with easily moved covers, so that they may be emptied at the proper intervals of time, and thus ensure the maintenance of the proper flue spaces.

The duty to be performed by the drying plates and ovens is of such a character as to cause destructive wear and tear; and they form in those works, where used, one of the largest items of structural reparation. The high temperature of the contents of the ovens, and their lowering when being emptied or filled, produces such expansion and contraction of the materials of which they are constructed, as to necessitate frequent repair and renewal. The plates again receive the slurry in a cold and wet state, inducing thereby the sudden change of temperature, and consequent destruction of the coverings.

We perhaps may be considered as over careful in mentioning that to work ovens of this character with efficiency there must be almost practically air-tight covering to the flues. When covered with the slurry they will be effectually so; but care should be taken that when clear they are free from holes whereby air could be drawn in and prejudicially affect the working of the ovens, and interfere with the economy of the flues.

CHAPTER XVI.

KILNS AND MODE OF BURNING.

In the progress which has been made in the manufacture of Portland cement during the last ten years but little effort has been exerted in the direction of the kiln. The original pattern first introduced by Aspdin still prevails in all its variety of form and modifications, yet it is curious that none of those imitations reach the excellence of the one at the parent cement manufactory at Wakefield. The chimney or dome of that kiln is of unusual height, and much resembles a glass furnace in appearance. This extreme height, while affording excellent facilities for burning the cement, meets also a difficulty in reference to the nuisance created by the discharged gases during the process of calcination. Since the establishment of these works, the usual result of increasing population has encroached on them, and indeed must eventually surround the factory, with the unavoidable consequence of complaints about smoke, &c. There can, therefore, be no extension of the manufactory, and it is just possible that sooner or later it will have to be removed altogether. This difficulty has also reached other works, and manufacturers should direct their attention to the construction of those kilns that will, if not entirely dispose of, at least mitigate the evil, and thereby render it less objectionable.

The accompanying sketch in Fig. 22 is an average sample of the ordinary type of Portland cement kiln, and it is surprising that no active measures have been taken to supersede this form for one based on a more scientific model. It is true that various modifications or improvements on this plan of kiln have from time to time been made, but they do not aim at any alteration in its form, but rather attempt to

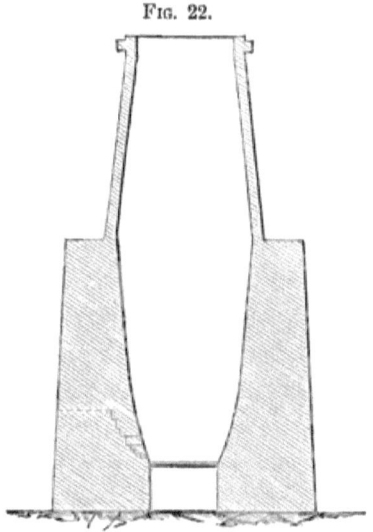

Fig. 22.

utilize the waste heat, and so partially abate the nuisance of the resulting gases. The earliest application of this kind was at Northfleet, where the kilns were closed after ignition, and the heat and gases evolved during burning were conveyed by a series of flues for the purpose of heating the drying floors, on which the slurry or dried raw material was placed on its being taken from the backs or reservoirs. This arrangement was regarded as successful, and at least saved the cost of separate drying floors heated by flues, but we do not regard it as at all economical or advantageous, as it interferes with the profitable and rapid exit of the carbonic and other gases, resulting in a variable quality of cement obtained at an increased cost of fuel. Indeed, this plan is only tolerable in a large factory where the contents of an inferior or irregularly burnt kiln can be blended with more accurately made cement, and its dangers and imperfections are thus mitigated, if not extinguished.

In Germany for many years the Hoffman type of ring or annular kiln has been used with much advantage, and perhaps would have received more extensive development but for the constant litigation attending its use, and the legal conflicts arising from attempts to evade the patent rights. Hoffman's patent in England has happily run its length, and the free use of the invention is no longer interfered with. We have had many of the so-called Hoffman's kilns built in this country for burning bricks and lime, resulting in much advantage not only in the quality of the manufactured article but in the cost of fuel required for its production. There has been much prejudice against this admirable form of kiln, and for lime-burning purposes great loss and anxiety have arisen through ignorant treatment and interested opposition, principally from the workmen. The extinguishment of the enthralling patent and its attendant impediments has cleared the way for free and healthy action in the direction of the ring kiln, and we may look forward to an early outburst of improvements in more reasonable and less expensive adoptions of its principle. It is somewhat singular that no Hoffman's kiln was used in this country for the purpose of cement making during the existence of the English patent, but now, and only very recently, one at least has been erected at the oldest works on the Thames. If a Hoffman's kiln was useful anywhere, it was essentially so in this country, where the fuel of any quality was easily obtained, and the quantity of cement made was sufficiently large to warrant a trial at all events of its doubtlessly valuable qualities. It is just possible, however, that some reluctance was felt in abandoning a system of kilns in which could be used the gas coke of London at a cost that under certain circumstances could not be reduced by any other mode of burning. Probably the cost of gas coke in the most favourable and cheapest times could not be bettered by even a Hoffman kiln in the simple direction of fuel cost ton for ton; but burning is not the only advantage offered by the ring kiln. In saving of labour there is much superiority over the old kiln, and the convenience of having the operations of your works performed at a uniform level is not the least of the advantages obtained through its agency. The certainty also of being able to utilize the waste heat as it radiates from the incandescent mass in the preparation of the adjoining compartment for the first stage of calcination is a great advantage.

Probably the best way to illustrate this important subject is first to consider the necessary conditions requiring accurate fulfilment in burning Portland cement. The preliminary process of selection of the raw materials and their perfect amalgamation must be regarded as a *sine quâ non*, otherwise it is needless entering into further processes of any kind, for the burning or grinding cannot cure the evils of imperfect proportions or admixture.

The required heat, through whatever agency it may be applied, is for the purpose of uniting in profitable combination the mixed raw materials, and thus forming an hydraulic or Portland cement.

The stages of this operation may be regarded as threefold, and are:

- First.—The expulsion of the water remaining in the bricks or dried slurry.
- Second.—The separation of the carbonic acid gas contained in the carbonate of lime proportion of the mixture.
- Third, and most important operation.—The combination by fusing or clinkering of the lime with the silica, alumina, protoxide of iron, and alkalies resulting in the desired product.

To perform this threefold process accurately it is necessary that each of these canons of manufacture should be regarded as separate and distinct, taking its place in due order. Thus, until the expulsion of the latent moisture, fire cannot profitably if at all expel the carbonic acid, and the increased heat for the clinkering should follow and not precede the decarbonating operation. The reactions take place at different temperatures, and provision should be made for a gradual raising of the heat in the order above referred to. It is most important, however, that the last and finishing stage in the application of heat for the clinkering should be capable of accurate adjustment. As the temperature required for this purpose is a high one (white heat), and converts the mass into a molten condition, care must be taken that it should neither last too long, nor be excessive in character, otherwise the product realized would be a glassy mass, incapable of being used for hydraulic purposes. In addition to this waste, the lining of the kiln would become fused and amalgamate with its contents, so that the act of drawing or emptying a kiln so improperly burned would entail great loss for structural repairs.

To fulfil the necessary requirements for the due and profitable performance of these unalterable rules, it would be requisite to have a kiln so arranged as to secure the various temperatures and their controlment. Let us examine, in furtherance of the elucidation of our subject, the general type of dome kilns (Fig. 22), and how far it is capable of fulfilling these conditions.

The treatment and use of dome kilns may be regarded as purely empirical in character, and therefore subject to the irregularities and vicissitudes incidental to an unguided system.

The operation and the conditions surrounding its performance are as follows:

A cooled and empty kiln is thus treated:—The bars are

duly placed in position, and on them are laid the fagots, or fuel of ignition, followed by a layer of coal or coke, succeeded alternately by dried bricks or material in the recognized order and quantities. There is no specified standard of weights or measure, at least not any of a trustworthy or exact character; for the condition of the fuel and materials renders such exactness impossible. The whole matter, therefore, is left to the experience of the burner and his assistants. The kiln is lighted, and the gases and products of combustion permeate into and through the superincumbent mass of coke and raw cement material, gradually expelling the moisture and rendering the contents of the kiln warm. Then the coke becomes ignited, and the walls of the kiln get hot. This, the first stage, can be observed by an experienced eye when the vapour has been expelled and all organic matter changed in character. The completion of the second stage arises when all the materials of combustion have become profitably heated, and the carbonic acid drawn off, which is about the temperature of a bright red-heat. The quantity of carbonic acid thus requiring expulsion is something enormous, so much so that for every ten tons of raw Portland cement, four tons at least of this acid were set free. So that when you estimate the weight of coke, or other fuel, and raw material put into a dome, or any other kind of kiln, and compare it with the cement produced, you can form a fair idea of the amount of power absorbed in its handling. Heat can only be regarded from a scientific point of view as so much power. Another loss of heat is effected through the carbonic acid, which carries off a considerable amount of warmth in a latent state. After the expulsion of the carbonic acid, the products of combustion become hotter, and the final or third stage of the operation begins. The economical and accurate attainment of this result can only take place when the raw material and the fuel are duly proportioned, otherwise it will either fail to produce the accurately clinkered mass, or destroy its value by producing the glassy and worthless clinker before referred to. The practised cement burner, by his "rule-of-thumb" process, estimates as nearly as his judgment can guide him the quantity of fuel required, and at the bottom of his kiln places larger or thicker layers of coke, and as he ascends puts thicker layers of brick or material, thus distributing as equally as he can the desired warmth. As the coke remains porous during the whole process, the contents of the kiln cannot clinker sufficiently to prevent the circulation of the necessary amount of air for combustion. Dusting of the contents of the kiln, however, either by excessive heat or imperfectly and inaccurately combined raw materials, checks the draught and retards to a wasteful length the completion of the burning. The clinker is thus produced in separate layers, and gradually cools from the bottom upwards, so that the layers are prevented from being clinkered into one mass and easily separated when the kiln is ready for drawing.

When an operation of the above character thus intuitively guided, and beyond the influence or controlment of rigid technical rules, succeeds, there can be no reasonable objection to its continuance where so loose and unintelligent a supervision exists as to accept its services. However, the waste of fuel is enormous, not only from the necessity of warming the cold kiln, but also from the amount of heat carried away through the agents of combustion, all of which is so much lost and unrecoverable. The mechanical and unavoidable destruction of the kiln forms another and somewhat serious item of expense from the inevitable expansion and contraction of the linings during the interval that occurs between its loading and unloading. These are, as well as can be shortly put, the conditions of the kilns used in

England for burning Portland cement, and forcibly illustrate the disadvantages which cement makers, who persist in their use, labour under. Nothing can be more wasteful or costly in character, unless it be the cement kilns described at page 171, and which were only tolerated, as there explained, under peculiar and exceptional circumstances. Cement makers surely are intelligent enough to realize the unhealthy and unbusiness-like condition in which they persist in remaining, indifferent to the progress in the science of their art, which it is undoubtedly their first duty to encourage and foster. While the customer pays (and he really does) for the extra cost incurred in the manufacture of the cement, or until a period of great competition arises, we need not expect much marked improvement in this or any other antiquated part of the process. With the exception named at page 241, no effort has been made in a scientific manner to effect any improvements in the kilns. There is indeed, we must not forget, the proposition of Mr. Johnson to adopt his patent kiln for the utilization of the waste heat, but there is nothing scientific in its character, as the short description given farther on will show.

The desire to take advantage of the waste heat and employ it profitably, while overcoming the difficulty of surrounding opposition on the score of nuisance, is a most laudable one, and deserves such encouragement as the merits of the invention warrant, but not to the sacrifice of other and more important advantages.

Having thus, we hope, given a clear description of the old and faulty type of cement kiln, with the inherent defects inseparable from its very constitution, we will proceed to examine the Hoffman's ring kiln as shown by Figs. 23 and 24, and point out the salient features whereby its claims to superiority are undoubtedly established, Fig. 23 showing sections of flues and chimneys with the valvular damper

KILNS AND MODE OF BURNING. 247

arrangements, and Fig. 24, a plan, partly in section, of the chambers and fire-holes.

Fig. 23.

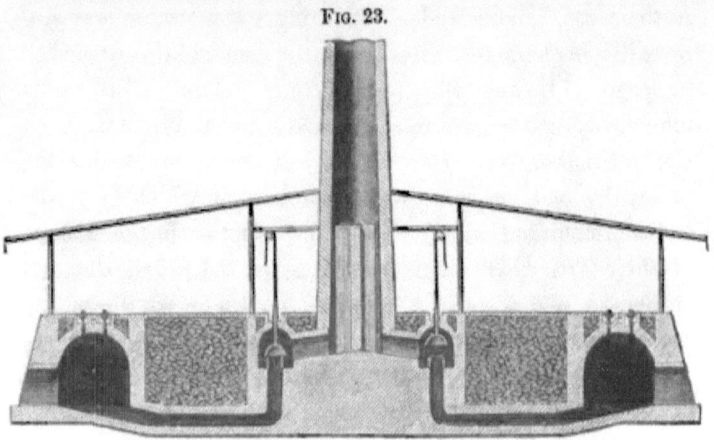

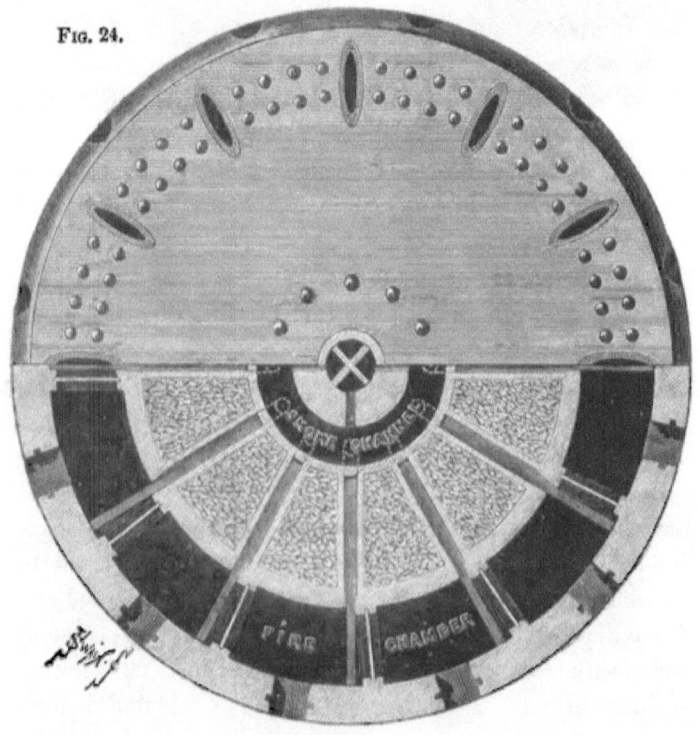

Fig. 24.

The essential features by which this kiln is distinguished consist in the construction of a series of variable chambers surrounding a high chimney, to which they are connected by properly arranged flues, having the necessary valvular or damper arrangement for their efficient and economical control. The primary objects attained through this scheme of chambers, flues, and shaft or chimney, are the conversion of the fuel into the various gases, their ultimate, as near as may be, perfect combustion, the evaporation of moisture, the expulsion of carbonic acid, and perfect vitrifaction. These processes come in regular order, and their successful completion depends on the accuracy or carefulness with which the bricks of raw material are built up, for as they perform the duty of generators for the gas, any carelessness in this direction will materially influence the quality of the results obtained. Each chamber is so constructed that it can readily be put in communication with the outside or atmospheric air, the central chimney, and adjoining chambers; the first to obtain the necessary air for combustion; the second for the purposes of draught, and the last to receive waste heat from one chamber, and in due course, as the combustion proceeds, to pass it on in turn to the next. The fuel is introduced at the top by properly arranged openings, so that the temperature can be efficiently controlled and regulated at pleasure. It will be observed that in this, as in the other kilns, it is necessary that the chamber burned off must cool (but its external walls are warm) before its contents can be dealt with, and the only and important difference is, that while by the old form the heat is lost and dissipated, the ring kiln permits and secures its almost perfect utilization.

The process of ring-kiln burning is a continuous one, and admits of no cessation, otherwise the best advantages to be obtained would be lost. This facility of, as it were, perpetual working is, however, accompanied by the disadvantage of great loss if there is not enough work to keep the kiln going regularly. That is to say, if the system cannot be conducted in its entirety, do not attempt to do it partially, for such a course would result in certain failure and dissatisfaction.

A marked feature in the working of the ring kiln is, that after placing the materials to be operated on in due and regular order—and this is important—you still hold in your hand the fuel, and can control its application, regulating its supply according to the circumstances by which you are surrounded. In fact, it is never lost sight of, for you can see its effect, and change at will the quantity, or control its results, by the mechanical arrangements which regulate the admission and exit of air from the flues. The working of this kiln entails a certain amount of technical skill, for its conductor should be able to determine the quality of fuel required and most suitable, as well as competent to judge of the various temperatures and the length of their duration. The facility with which observations can be made during the progress of burning forms a marked contrast to the conditions attending the operation of burning the dome kiln, for there you commit your fuel to the mercy or accident of changing weather, and must stand by utterly helpless to remedy any defect, however trivial. You have exhausted your ingenuity when the kiln is shut up, and the fire started, and you cannot estimate the result of your endeavours until the kiln is again cold, and its contents ready for grinding. It would be unfair if we were to disregard or fail to acknowledge the amount of careful and effective work performed by experienced burners, controlled by equally intelligent foremen, in the large works on the Thames and Medway, and the criticisms made on the existing system are merely intended to call attention to this subject for the sake of comparison.

There is a weak point even in this useful ring kiln, and one that does not receive its development in the burning of bricks or lime, because these materials maintain their normal form during the process of burning, unless under circumstances of great carelessness. The mixture of raw materials required for the successful fabrication of a good Portland cement is primarily designed to promote vitrifaction. This needful and most important result is attended by a changing form of the bricks, and the tendency by this change to destroy or lessen at least their value as gas generators or regenerators. The desired object and its most commendable performance is, when the bricks are, though contracted, not much reduced in size, and competent to retain their useful regenerator form, and at the same time clinkered enough for the production of a good Portland cement. In the best conducted cement works where the ring kiln is employed, the raw bricks are so built that a grate-like structure results, thus facilitating the passage of the air and heat equally through the whole mass or fabric and its exposed surfaces. In the treatment of this kiln it is preferable that the upper bricks should be slightly under-burnt, so as not to endanger the economy of the finishing or clinkering operation by allowing the foundation or substructure of the contents to be deranged, and thus causing imperfect draught and circulation of the air and gases. The observations we have offered on this kiln and its claims to superiority over the common ones, are intended, in addition to their explanatory character, to show that a kiln of this scientific design cannot be successful unless built and used on the lines so accurately laid down by Hoffman and his agents. Any departure from its prescribed form or interference with the flues or valvular arrangement is pretty sure to result in disappointment. In this as in every other kiln of whatever form it is indispensable that the cement mixture should be unvarying and exact.

Some years ago much litigation occurred in Germany owing to the misunderstanding about a Hoffman kiln erected at a Portland cement factory under the supervision of Hoffman or experts in his employ. Failing to burn good cement, the proprietors of the works claimed damages from Hoffman, which he resisted, and the matter coming before the usual tribunal for settlement, much useful technical evidence was forthcoming on both sides; the contention on one side being that under no circumstances was it possible to burn a good normal Portland cement in the kiln in question, and on the other that not only was this possible, but in numerous cases it was actually being performed in various districts where all kinds of raw material were being used.

In the result it was satisfactorily shown that the cause of failure arose from a combination of circumstances, more especially by reason of imperfect reduction of the raw materials and inaccuracy in their relative proportions. The careful examination of the faulty products furnished conclusive evidence of slovenly or ignorant management, and successfully established the advantages of the ring kiln when controlled by intelligent supervision. The imperfect grinding prevented the necessary close contact of the raw particles, thus rendering impossible the accurate chemical result which is only attainable, as we have elsewhere shown, when the raw powder or dust is almost impalpable.

During the examination of the various processes in preparation for this trial, one of the witnesses, in furtherance of his views, instituted a series of experiments to show that as good cement could be made from a ring kiln as from the old dome kiln. To give weight and value to these tests, he procured a sample of the best English Portland cement, made by the first makers in a dome kiln, and compared it with a cement of high repute made in Germany, and obtained through the agency of a Hoffman or ring kiln.

The analyses of these two cements were as follows:

No. 1.—English.
,, 2.—German.

	1.	2.
Silica	22·74	21·11
Alumina	7·74	11·30
Oxide of iron	3·70	3·36
Lime	56·68	58·03
Magnesia	0·57	2·93
Alkalies	0·63	0·71
Sulphate of lime	1·66	0·51
Carbonic acid	3·50	0·83
Undissolved residue	0·53	0·49
Water	1·90	0·54

These cements were tested in the following manner with different proportions of sand and at various intervals, and like the above analyses indicated a very even quality, showing that they had been carefully made; and when a cubic foot of each was accurately placed in the measure and then weighed, the following result was obtained:

	Lbs.
English cement	105·40
German	106·37

The materials used in making the English cement were white chalk and Medway clay, or mud; and those for the German cement were limestone and clay; the English cement being burnt with gas coke and the German cement with coal. The tests were carefully conducted, and the recorded breakings were as follows:

The moulds used were made of iron, and gave the usual section of 2·25 square inches on the breaking surface.

English Portland Cement Mortar.

Mixture.	Age.	Tensile Breaking.
	days.	lbs.
1 cement and 1 sand	8	283·4
,, ,,	16	371·1
,, ,,	24	580·3
1 cement and 2 sand	8	191·1
,, ,,	16	236·1
,, ,,	24	390·3
1 cement and 3 sand	8	103·4
,, ,,	16	179·4
,, ,,	24	277·3

German Portland Cement Mortar.

Mixture.	Age.	Tensile Breaking.
1 cement and 1 sand	8	230·4
,, ,,	16	381·4
,, ,,	24	601·3
1 cement and 2 sand	8	173·7
,, ,,	16	245·3
,, ,,	24	411·4
1 cement and 3 sand	8	98·3
,, ,,	16	191·4
,, ,,	24	290·3

To show that even with the best of raw materials found on the ground, and a Hoffman's kiln, the owners of the works in question failed to produce a good and marketable cement at the time referred to, we give the analyses of chalk and clay:—

1st.—Chalk.
(Three specimens.)

	1.	2.	3.
Carbonate of lime	98·59	97·89	98·11
,, magnesia	0·13	0·21	0·19
Oxide of iron	0·34	0·33	0·37
Silica	0·92	1·56	0·83

The Clay.

Silica	64·11
Alumina	20·77
Oxide of iron	8·43
Magnesia	1·63
Lime	1·01
Potash	1·84
Soda	1·24

The proper mixtures required from the above excellent materials would, after allowing for the silica, which was rather flinty in character, produce a cement of the following analysis:

Lime	60·0
Silica	25·0
Alumina and oxide of iron	12·0
Magnesia	1·0
Alkalies	1·0
Residue, &c...	1·0

The above analyses and tests were made in the most careful manner by Dr. Fiureck, of Berlin.

There is no doubt that now, with the experience and loss sustained, principally through careless treatment and utter disregard of the technical department, the cement at these works is of good and reliable quality.

The Hoffman kiln, constructed on purely scientific principles, requires a more exact system of management, and in many cases a special treatment of the raw materials and their proportions. A mixture producing good cement in the common kiln would probably not be equally good if burnt in a Hoffman kiln. Again, much care is necessary to prevent over-burning in the ring kilns, so as not to produce a slag or glassy mass. Such a result destroys the circulation, and the heat ascends above the contracted and sunken mass, thereby impinging in all its intensity on the arch, which would, under such circumstances, be destroyed.

In somewhat close imitation of Hoffman's is the kiln suggested by M. Lipowitz, which, instead of an expensive main chimney, provides for draught through the agency of ventilators. In the ultimate disposal of the resulting heat, he, by an arrangement of air-channels, dries the cement brick after being formed at the brick press. This kiln, on its first introduction, too closely imitated the design of Hoffman, and in consequence, during the existence of that patent, a certain amount of risk attended its building or use. There are many points of excellence in this scheme of kiln, and in part or whole it may some day be usefully applied for burning cement.

A novel and ingenious system of burning bricks, lime, cement, &c., has been devised by M. Bock, and its merits are worthy of a slight description. (See Figs. 25, 26, and 27.)

This kiln consists of a long horizontal channel, 1 metre broad and 1·3 metres high, built of fire and common bricks with strong iron ties, and is provided with air-insulating spaces, as is also the chimney, which is 20 metres in height.

Fig. 25 represents a longitudinal view of the kiln, chamber, shaft, firing holes C, and transit or wagon channel D.

Fig. 26 is an enlarged view of the axle and wheel arrangement of the wagons, and the mode by which, in their passage, they are rendered air-tight.

Fig. 27 shows section of kiln.

The materials to be burnt are placed on specially designed trucks, which run on rails fixed in the bottom of the channel, and are pushed on to the firing holes, situated about half-way in the channel. An arrangement of guides at either side of the wagons dips into a sand trough, insulating, as it were, the metal wheels and axles from the heat of the firing channel proper. The section of the channel is varied according to the object for which the kiln is to be used. It is found that for bricks a square form is sufficient, but for lime and cement it is necessary to build the walls with a batter, and the internal section is also narrowed towards the top, with stronger arches to resist more effectually the higher temperature necessary for these purposes. To enable the cast-iron platforms of the trucks to resist the high temperature, they are covered with a double layer of fire-bricks bedded in refractory clay and plastered on the surface as well. By this arrangement, and the sand troughs in which run the side guides of the trucks, the channel is practically split into two horizontal divisions. The one, while serving

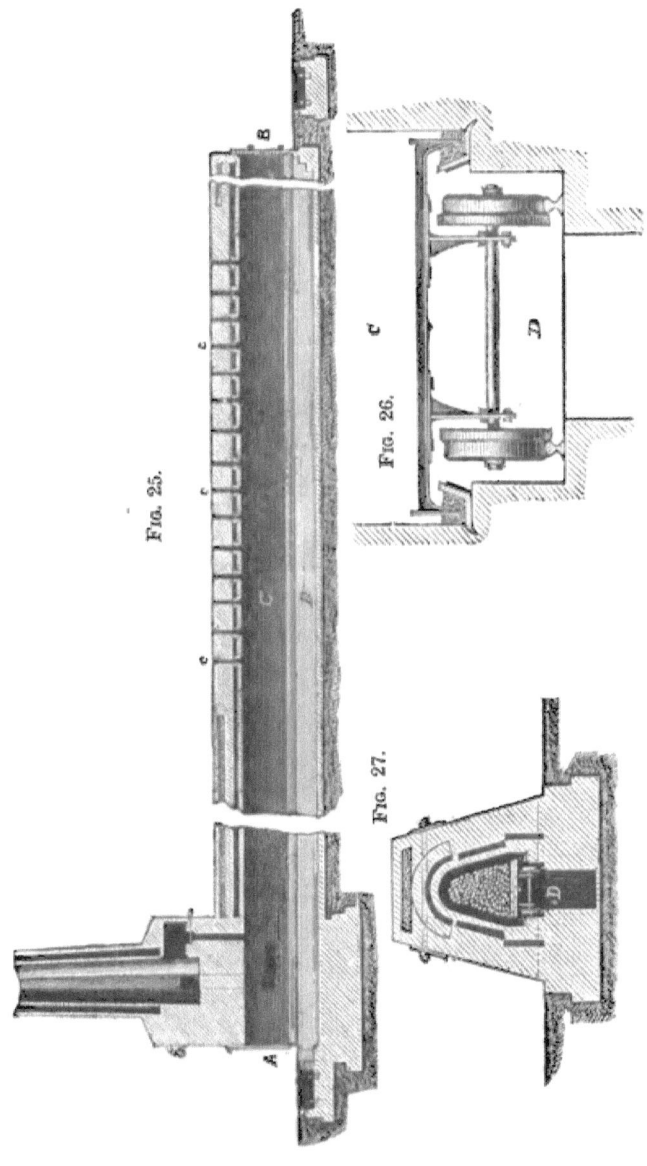

as the means of transporting the loaded trucks, also acts as a passage for the introduction of air for the purposes of combustion.

The chimney of the kiln is fixed at what may be called the loading or feeding end A, by which the loaded trucks are introduced. These trucks are in turn pushed along the line of rails, being accurately coupled to each other, so as to maintain the necessary separation of the rail and wheel and axle channel from the fire-way. Until the channel is completely and accurately filled the burning cannot proceed, for the splitting of the indraught of air from the returning or fire division of the channel draught would be incomplete. The charge, therefore, having been introduced, both ends of the kiln are closed with strong double iron doors accurately fitted together.

The firing is now begun at the fire-holes, and the fuel introduced in a somewhat similar manner to that adopted in the ring kiln. The necessary air is admitted at the entrance, passes under the wheels and axles, keeping them cool, and at the closed doors of the end B returns over and through the contents of the trucks. The heat can be controlled and regulated by the valvular arrangement at the bottom of the shaft in conjunction with the supply of air, and accurate observation can at all times be made through the holes of fuel supply, which are properly capped. Each truck is loaded with 500 bricks or its equivalent in weight of other materials, such as lime or cement. The original method by which the trucks were propelled was by a strong screw which pushed a truck its full length within the channel at the entrance A, and at the exit B a truck emerged with its load burnt, and was then again ready for further loading. The kiln is therefore continuous in every sense, and when a larger capacity is desired it is only necessary to lengthen the channel or duplicate it.

In the experience which has been obtained of this kiln, some alterations and improvements have been found desirable. The stoke or fire holes were found to be insufficient, and they are now assisted or replaced by side gratings. The cast iron of the trucks was changed for wrought iron, as it was found that the former fractured with the high heat. The lubricating agent used for the trucks is a mixture of graphite and tallow, it being competent to resist the high temperature. The screw for pushing the trucks was found inefficient, and an hydraulic press arrangement was substituted; the level channel was then made 1 in 100, favouring thereby the passage of the trucks. There is still some difficulty owing to the small pieces of coal or cinder getting into the sand recess, and interfering with the free passage of the trucks.

The cost of these kilns in Germany is as follows. Brick-burning kilns,

			£
To burn 6,000 bricks daily	..	.. ..	525
,, 12,000 ,, ,,	..	.. ..	675
,, 24,000 ,, ,,	..	.. ..	1125

The number of bricks required to build those kilns are:

		Bricks.
For a make of 6,000 daily	.. .. ..	120,000
,, ,, 12,000 ,,	.. .. ..	150,000
,, ,, 24,010 ,,	.. .. ..	200,000

A Hoffman kiln to burn 8000 bricks daily, requires 450,000 bricks.

These kilns are, so far as we are aware, used only for brick burning, but the principle with its manifest advantages could with some alterations be used for lime and cement burning.

Although the name Bock is given to this kiln, its real inventor is supposed to have been M. P. Borne, of Paris, who upwards of twenty years ago recommended a kiln on this principle of construction.

The accurate and profitable operation of any of these scientifically arranged kilns is prejudicially interfered with if the technical conditions on which their principles are based is departed from. Although, theoretically, the unburnt materials are supposed before being subjected to the beneficial action of the heat to be thoroughly dried, and all the latent and acquired moisture of plasticity eliminated, it is seldom in practice that we find these conditions complied with. The consequence of such carelessness is to hinder reliable observations during the early stage of calcination, and in some cases to prevent altogether the realization of the advantages of the kiln. In using the old style of kiln the materials when placed are seldom thoroughly dry, and the coke, under the most favourable circumstances, contains from 20 to 30 per cent. of moisture, all of which has to be evaporated before any profitable action can take place. The drying in the kiln becomes therefore a necessity, and follows as a matter of course. In the ring kiln, however, no such necessity arises, and indeed the very fact of steam being emitted through the chimney indicates not only a slovenly or careless treatment, but an utter ignorance of the principle of the kiln itself, and the rules by which its true working should be governed. It is true that one of the leading features in the ring kiln is its capacity for the utilization of the heat from the burning mass, which heat, after the operation of burning has been performed, is passed into the adjoining chamber to assist in the preparation of its contents for the finishing process; so that this kiln arrangement is designed for burning a previously desiccated mass, or at all events one deprived of all superfluous moisture. The kiln is not to be used as a drying oven nor to receive the moulded bricks from the forming machine. It has sufficient work to do in facilitating the decomposition of the carbonate of lime, and its economical conversion and combustion of all the contained gases. The indraught of air required for the purposes of combustion is effected through the agency of the chimney, and little if any allowance is made for a body of steam, whose existence in large quantities would only act as a conductor and rob the heat from the kiln. A ring kiln, therefore, which exhibits a cloud of steam from its chimney, may be regarded as one under the control of inexperienced or ignorant management. In a small kiln of this type it may not be quite possible to extract from it the full value of which it is capable, but in those large enough to permit of its fullest development, the practical result should be perfect combustion, and the air entering the chamber of the shaft should not exceed 120° in temperature.

If carelessly erected, and on such foundations as render the extraction of moisture from the ground possible, the behaviour of the kiln will always be unsatisfactory. In well-arranged kilns not only are their foundations so constructed as to render impossible the absorption of damp from the ground on which they are built, but the chamber and other exposed external portions are carefully clothed by sufficient covering so as to prevent the radiation of heat. Indeed the machinery of the kiln (for its flues and their controlment may be regarded as mechanical) is primarily devised to consume with accuracy the required amount of fuel according to the purpose for which it is designed, and in its combustion perform this important duty.

Johnson's improved Portland cement kiln differs but slightly in principle from the old closed kiln, by which the waste heat was utilized through the agency of the draught obtained by a chimney of great height. By this process—for much of the old practice is abandoned by the adoption of this kiln—the usual method of washing with harrows in the circular wash-mill is resorted to, and the washed or mixed materials pumped up to a level above the kilns into cylindrical reservoirs. After a certain amount of decantation these reservoirs empty themselves by the removal of the slucies on to the drying channel, warmed by the gases in their passage to the high chimney.

In the extensive works recently erected (on his plan) by Mr. Johnson, near Greenhithe, a marked difference in the extent of the back-room is apparent. There are a couple of small backs apparently used as reserves, but the space occupied by the circular receivers is very inconsiderable indeed, and must reduce the cost of manufacture to a marked extent.

Under any circumstances and by whatever agency the process of burning, as it is usually termed, is performed, all aim at one and the same object, viz. the production of a clinkered material of the following characteristics and properties:

1st. In appearance it should resemble lava in texture, and be more or less porous according to the materials used, their treatment, and the amount and value of the applied heat. The object should be to produce a clinker of a greenish or bronze tint, and not of a dark blue, which, if produced, indicates a too dense mass, incapable of energetic action, and when ground produces an angular powder of the most sluggish binding capacity.

2nd. It should have the power of absorbing, when finely ground, a moderate amount of water, setting within a reasonable time, and resisting when set the action of water. The specific gravity should not be under $2 \cdot 80$ or exceed $3 \cdot 2$.

CHAPTER XVII.

BRICK-FORMING MACHINERY.

In the dry process it is necessary, under the best application of that system, to convert the powdered materials into convenient forms for further treatment. The first duty of the cement maker is to prepare the dry material by the addition of water to produce such a state of plasticity as will suffice to permit of its being pressed or formed in the brick machine selected for that purpose. This is the place where it will be most convenient to impart to the mass any chemical elements of which it may be found deficient, so as to assist in securing the required clinker from the kiln. The production of a first-class Portland cement is only possible when the combined ingredients are competent to perfect the chemical result.

In the absence of the necessary alkalies, a dusty or disentegrating clinker will be produced, devoid of the required cementitious properties. The greatest care must be exercised, while imparting the requisite moisture to the powder, that no mechanical disarrangement of its atoms will result therefrom. The process of reduction will have brought the raw material into a fairly accurate combination, and all that is now required is to maintain that accuracy undisturbed. The primary object of the imparted moisture is to render the combined mass cohesive enough to ensure its passage to the kiln, and permit its being able to withstand the pressure to which it may be subjected while undergoing the process of calcination. However small the amount of water used, it may be regarded simply as an agent by which the powder is rendered plastic, unless where any chemical agent is dissolved in it for the purpose before referred to. The water therefore, is practically a temporary expedient, and it must again be passed off in a state of vapour before the bricks can be properly acted on by the fire. The primary object of this process is to facilitate the passage of the ground materials to the kiln.

There are various kinds of brick-forming machines, and as true accuracy of form is unnecessary, there is no need to use a very expensive or needlessly complicated one. In Fig. 28 is shown a very useful pug-mill machine, competent to exert the required pressure on the plastic mass and pass it on in a fit state to the drying sheds.

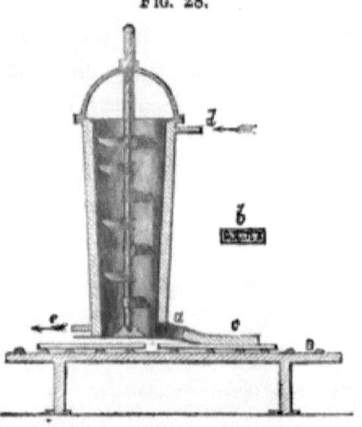

FIG. 28.

This brick machine is of the simplest kind, and can be made of any required size. The dust is received from a large hopper, or room, situated at a level high enough to permit of the dust traversing a sufficient distance between double screws or creepers, during which time it receives the necessary amount of moisture. On entering at the top of the pug-mill it is acted upon by the knives attached to the spindle, and pressed down to the bottom of the mill, where it is extruded at *a*, forming a continuous bar to *o*, and is then passed on, divided or cut as required, by the rollers of the receiving frame to B, from which it is transferred to the proper wagons or barrows. If required, the machine may be made to deliver at both sides. To facilitate the drying of the mass, the mill is surrounded by a steam or watertight jacket, where steam or hot water can be introduced at *d*, and after surrounding the pug-mill, finds an outlet at *e*. I has been proved by experiment that if the water is applied hot to the powder the time required for drying is much reduced. There is some risk, however, in too suddenly depriving the mass of its imparted moisture, and care should be exercised when this plan is adopted. The form of the brick as represented by *b*, need not be regarded as an arbitrary one, but capable of adaptation to the wants and requirements of the manufactory. Some materials, especially those of a crystalline character, even after the most exhaustive and careful treatment, develop a shortness, or granular character, and cannot readily be made compact enough under the ordinary treatment, to prevent crumbling and dusting of the bricks during their passage from the brick machine to the kiln. Under such conditions the moisture should be retained as long as possible, and, indeed, in some cases twenty-four hours' soakage of this powder would be advisable before being formed. Not only is the close contact of the prepared raw material desirable, but their true accuracy should be secured until the application of the kiln heat. Dusting of the raw material in the kiln is attended with injurious results, as the disintegrated dust of the bricks enters into and clogs the air-spaces required for the proper and economical combustion of the fuel. It is this difficulty which has, in several cases known to the author, resulted in failure, and disbelief of the impatient in the possibility of converting hard limestones and indurated shales into Portland cement. We must admit that there are difficulties attending the dry process under such circumstances, and especially where the necessary conditions of accurate manipulation are absent. It is impossible, however, to provide against ignorant and careless treatment in such a manufacture, although improved information and specially adapted machinery may, and does, sometimes assist the most undeserving.

Much of this difficulty may be overcome by a machine capable of exerting great pressure on a comparatively dry powder. Hitherto all machines for this purpose have more or less dealt with the material to be converted in a variable state of plasticity, rendering unavoidable a large percentage of water and air present in the pressed brick. In some clays used in the manufacture of building bricks this porosity of the product is not regarded as either objectionable or dangerous. In other qualities of clays, however, the presence of so much void interstitial space renders the brick worthless in character and dangerous to use where stability of structure is necessary or desirable.

As we have endeavoured to show that a highly compressed state of the raw powder is of the utmost importance, we will shortly describe two useful machines well suited for this purpose.

Fairburn's Direct-acting Steam Press.

The desired advantages for the perfect amalgamation of the raw material powder can be reached through this machine. In our desire to ascertain the latest and most approved press from a cement-making point of view we have examined the various presses and machines, and the result of these inquiries is that we regard this simple machine as very suitable for the purpose.

It is actuated by the steam pressure direct from the boiler, and therefore dispenses with all expensive and dangerous gearing, shafting and belting being avoided. Hence in a cement manufactory, where there are already too many mechanical complications, this machine will be found convenient and useful. We shall proceed to give a general description of it as shown in Fig. 29.

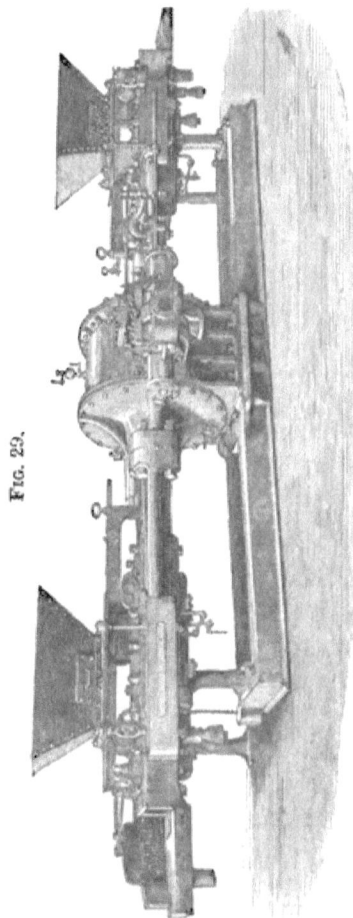

Fig. 29.

The machinery of moulding is actuated by the piston, at either end of the cylinder, and is, as shown, duplicated at each side thereof. The hoppers, which are so placed as to receive the previously prepared powder, are filled by elevators or other convenient machinery, or the supply can be accumulated in a store or floor directly over the machine.

Immediately below the hopper, and at the aperture through which the powder or dust passes, is fixed an accurately arranged measuring box, working over a slot on the upper side of the moulding die or tubes. This box is capable of easy and accurate adjustment, passing the required quantity according to the size and form of the desired block. This measured quantity drops readily through into the tube in which it ultimately receives the necessary pressure. On first exerting the steam power by the pressure from the boiler, the piston of the cylinder after being filled with steam moves in the required direction, and in doing so brings the box containing the measured quantity into a position for dropping its contents into the tube, as already stated. While this feeding action is proceeding on one side, the process of compressing and delivering is taking place on the other. This operation continues until the traverse of the piston has been exhausted or performed, and the desired density of the moulded mass is completed. The peculiar arrangement for effecting this object is of so accurate and perfect a character, and always so under the control of the attendant, that at will the pressure may be increased or diminished.

In the manufacture of bricks or blocks requiring to be of special form, this is readily effected by the introduction of templates or pallets giving forms of all kinds, either rectangular or spherical. In the latter case a most desirable improvement could be effected in cement making, for the balls thus made would be so dense in character and dry as to permit of their being at once placed in the kiln. Indeed they might roll into the kiln from the machine.

The steam pressure required to work this machine should be at least 50 lbs. per square inch, and where convenient higher pressure would be found more economical.

The output from a machine of this class working at a pressure as above would, supposing the balls to weigh 5 lbs. each, produce about 50 tons per day. One of the least costly of those machines having a 20-inch cylinder would exert a pressure of about 8 tons on each ball, and of course by increasing the pressure of the steam the ultimate possible pressure is almost unlimited. The expulsion of air from the balls or bricks is readily accomplished through the agency of the pallet boards, which can be made of a perforated or cellular form. The peculiar action of this machine is not spasmodic in character, but performs an even and continuous pressure during the passage of the moulded block through the entire length of the tube in which the brick must necessarily travel before reaching the point of delivery.

At first sight, without further explanation it would appear that the moulding of each block required the exertion of the filled cylinder, and only one block could therefore be made at each stroke of the piston. This is not so, however, as only a portion of the steam passes, leaving behind the remainder, which continues to repeat the strokes as desired, and according to the size of the block even as many as twenty at a time can be produced by each die.

Fig. 30 represents a perspective view of the machine known as

Guthrie's Brick-making Machine.

Hitherto much difficulty has been experienced in obtaining a brick-forming machine or press, capable of treating dry materials with any degree of success. In the increasing desire for a machine of this class many inventions have been offered to meet the required desideratum, but they have generally fallen short in being only capable of treating the clay or mixture either in a semi-plastic state or at all events too moist for economical or successful cement-making operations.

A machine, apparently possessing the necessary conditions of success, has been recently patented, and from the examination and performances of the machine submitted to the author's inspection, there is every indication of its proving a useful auxiliary in the manufacture not only of cement, but also in the conversion of hard and obdurate shales and clays into bricks.

Figs. 31 and 32 represent the machines as most recently improved, Fig. 31 being the elevation, and Fig. 32 the section.

The previously prepared materials are conveyed to the hopper H in the most convenient manner by elevators or

Fig. 30.

other similar machinery, from which they descend by their own gravity to the measuring box B, the supply to which is regulated and adjusted by a small slide at the bottom. At h the hopper has an opening whereby at all times inspection can be made of the progress of the materials, as they descend, and before being subjected to the action of the

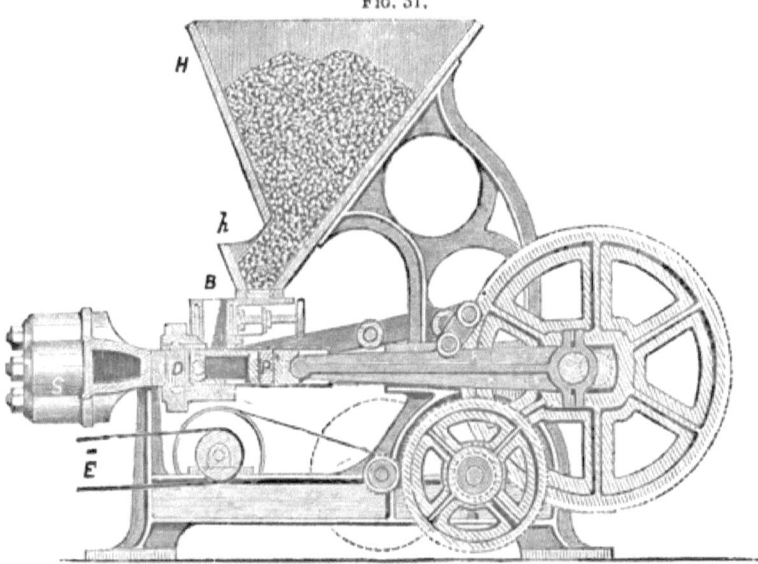

Fig. 31.

machine. The measuring box (B) has one adjustable side, so that the exact capacity may be readily adapted and regulated, moving on the mould M, to which it is fixed. D is the fixed or standing die against which the materials are pressed by the action of the piston or plunger P. To overcome the shock of the plunger, D is fixed upon a spring bed SS. The plunger P receives motion from a simple connecting rod and single-throw crank driven by spur gearing in the usual or ordinary way. A small pulley is fixed upon the pinion shaft which drives the delivering band E, upon which are received the bricks or blocks, when pressed and finished, on their dropping from the mould. The action or motion of the mould is obtained in a simple and peculiar manner, thus:—From a point about midway on the connecting rod, there is an elliptical movement which is attached to the rocking shaft above it, by a rod and lever, and by this arrangement great variation of speed is obtained. In addition, a short secondary stroke is imparted to the mould by causing the lever and rod (communicating motion to the mould) to pass the centre and produce a short and long reciprocation for every revolution. The object of this

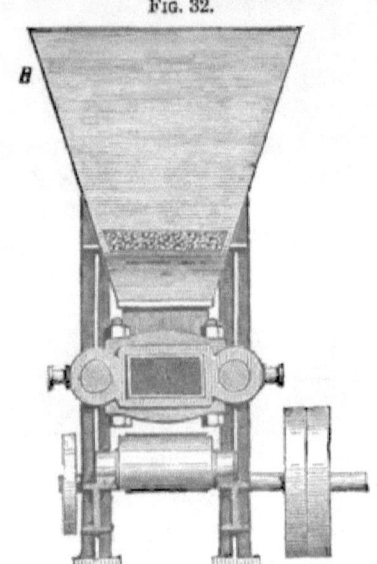

Fig. 32.

peculiar motion is to obtain accurate and sound corners to the blocks or bricks; the material being by this arrangement driven in both directions, while being submitted to the maximum pressure of the machine.

To overcome the injurious action of the air contained in the material under treatment two provisions are adopted. First. When the materials are introduced into the measuring box, they are permeated with steam, and then at once dropped into the mould, where the pressure is gradually applied before being completely closed up, thus allowing the dispersion of the greater part of the air or vapour before the maximum pressure is applied. Second. The arrangement of springs in connection with the die D, having a constant pressure of many tons, the greater part of which is maintained upon the block on its four exposed sides until it is entirely free of the mould, so that the atmosphere disperses any residue of air or vapour remaining, thus securing a properly compressed brick free from cracks.

Blocks made by this process have been tested to a pressure equivalent to a column of 900 diameters high. The edges of the block thus made have maintained their sharpness so much so, as to remain uninjured after being thrown about or pitched from a considerable height.

In some experiments made in presence of the author with a finely ground cement raw material the following results were arrived at:

1st. The required quantity of water imparted to a dry powder was only 15 per cent.

2nd. The capacity of the box B was equal to $177\cdot4$ cubic inches.

3rd. The brick produced under not by any means the extreme pressure which the machine appeared capable of expending was $98\cdot86$ cubical inches, being a reduction from the capacity of the measuring box B of 80 per cent. Weight of wet brick, $7\cdot87$ lbs.; when dry, 7 lbs.

Should it be found that the advantages promised by this machine are realized, and so far the author's examination confirms the promise made by its inventor—we shall have a new and powerful auxiliary in the dry process of converting the most obdurate materials into Portland cement. There is a possibility, however, that bricks so dense in character as those produced by this machine will require a larger amount of fuel for their conversion into cement, but this will be attended with the advantage of less waste in the kiln, and secure a resulting clinker of more reliable quality. The machine in question is competent by a varied arrangement of dies to produce any form of brick, and as in practice it may be found desirable to substitute some other than the square or rectangular shape, the facilities it affords for such a purpose are not the least of its advantages.

The bricks, made under the author's direction and from raw materials furnished by him, were so dense in character that for the purpose of examining their internal condition it was necessary to use a saw to divide them. Although the pressure applied during the experiment was much under the maximum power of the machine, yet the brick on examination proved to be so dense that it was difficult to detect the presence of voids or interstitial spaces. Under certain arrangements it appears to us that the bricks produced through this agency might at once pass on to the kilns.

Although thus presenting to our readers what we regard as useful machines for producing bricks, we do not insist upon their adoption; for any other of the numerous brick-making machines are more or less suitable for the conversion of the raw materials into the required form for further treatment. It is the duty, however, of cement makers to take advantage of the latest machines required for their operations, at least when the improvements tend not only to lessen the cost, but improve the quality of cement.

CHAPTER XVIII.

TESTING MACHINERY.

BETWEEN the simple and easily understood water test of Smeaton in 1757 and the elaborate scientific machine of to-day, there is a wide difference. It will be profitable, however, for us, we consider, to examine the various means by which the experimenters arrived at the respective results of their several investigations.

Smeaton merely tested the capacity of his mortar to set under water, and had no idea of measuring its cohesive or adhesive value. The binding properties which it exhibited satisfied his desire to have a reliable mortar with full capacity to hold together the materials of construction, and to withstand the disturbing influence of water action. The mode pursued by him is very similar to that frequently adopted at the present time of arriving at an approximate and hurried estimate of cement by mixing up a small portion and making a pat or cake only; instead of this form, however, Smeaton and Pasley made their mortar up into the shape of round balls.

Vicat, the first to associate with his investigations chemical knowledge to assist him in his experiments and the deductions he arrived at, went a step farther, and instituted what he called the needle test, performed by the aid of the machine shown in Fig. 33, being a representation of the original machine used by him.

We shall take Vicat's description of Fig. 33, which is as follows:

"The averge dimension of the different parts of the

machine is about an inch each way; breadth 10 inches, height from the sole to the lifting pulley, 21½ inches.

"Length of the rod $a\,b$, viz.: from a to c, 6 inches, from c to d, for the part enclosed within the cylinder of lead, 1·69 inches, from d to b, to the adjustment of the point p, 5½ inches; section ¼ inch (nearly); length of lift 3·937 inches.

"To make use of this machine, we commence by setting it up perpendicularly; the rod $a\,b$ being kept vertical. The cement to be tried is placed underneath this rod, the vessel containing it being wedged up if necessary; the point of the rod then bears on the surface of the cement. We read off at e, on the edge of the bent index, the number of tenths of an inch marked on the scale. We then lift it by means of the string f to a given height (fixed at 1·9685 inches for all our experiments), after which we release the string suddenly, like the monkey of a pile-driving machine. The point falls and penetrates more or less into the cement. We read off the scale a second time, and by subtraction arrive at the quantity of *penetration*.

"The rod armed with its point weighs 15383·7 grains, or 2 lbs. 3 oz. 2¼ drs. avoirdupois nearly."

All the compounds to be experimented upon were kneaded stiff, and as nearly as possible to the same degree of consistency, by the aid of a pestle. Those samples intended for immersion in water were put into rather deep than broad cups of delft or common glazed earthenware, and were covered with pure water immediately after their preparation.

The initial set of the immerged samples were measured by the number of days which elapsed from the instant of immersion to the time when they could bear without any appreciable penetration a knitting needle of ·047 inch diameter, filed to a right angle at one of its extremities, and loaded at the other with a weight of 10 oz. 9 drs. avoirdupois.

276   SCIENCE AND ART OF PORTLAND CEMENT.

The machine used for breaking the prisms of mortar, as shown in Figs. 34 and 35, is very simple in character, and

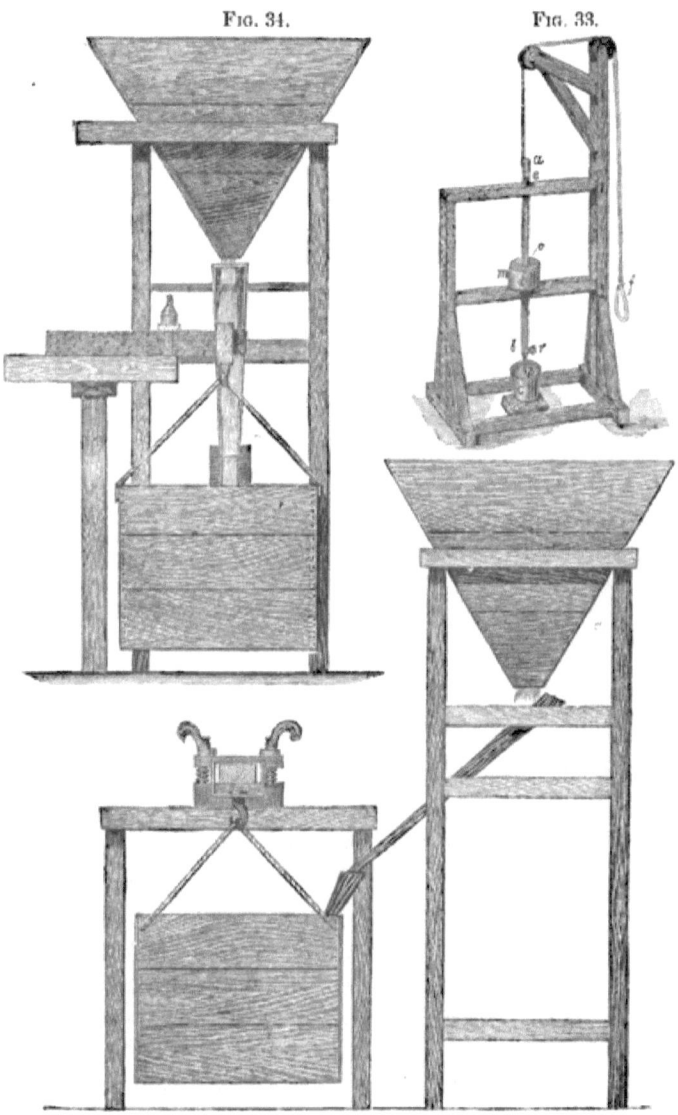

Fig. 34.   Fig. 33.

Fig. 35.

capable of accurately estimating the strength in the manner required. The machine must have been steady from the mode of applying the weight by a stream of sand, the passage of which from the hopper to the box suspended from the clip round the sample was even and smooth. Fig. 36 represents a modified form of Vicat's penetrating testing machine, which has been used in America and this country by some engineers and architects for many years.

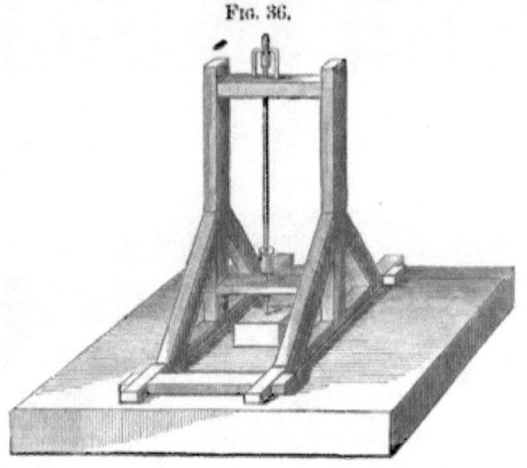

Fig. 36.

Pasley instituted a variety of tests of a practical character, embracing a wide range of investigation. He not only aimed at theoretical results on cements and limes, but extended his observations to their various combinations with building materials of every description. A favourite mode adopted by him and still practised by some makers was to extend from a wall a series of bricks joined together at certain prescribed intervals of time by the mortar under examination, and the number of bricks so cemented, and thus supported, indicated the adhesive value of the cementing agent. Such a test was of fluctuating and uncertain value, and no longer enters into the practice of modern testing.

278 SCIENCE AND ART OF PORTLAND CEMENT.

There were many other similar modes by which Pasley aimed at imparting to the constructive profession reliable knowledge in the most direct form. Those experiments on the strength of brick beams built with different kinds of cement mortar indicated his appreciation of that kind of structure; but modern or existing practice disregards such adaptation, and in such form is generally superseded by the use of iron. At the Exhibition of 1851 this form of test was adopted in the trial of Portland cements, but the results arrived at lost much of their value from the fact of iron hoop bond having been introduced into the beam.

Fig. 37 represents the apparatus first used by Treussart, and through the agency of which Pasley obtained his tensile

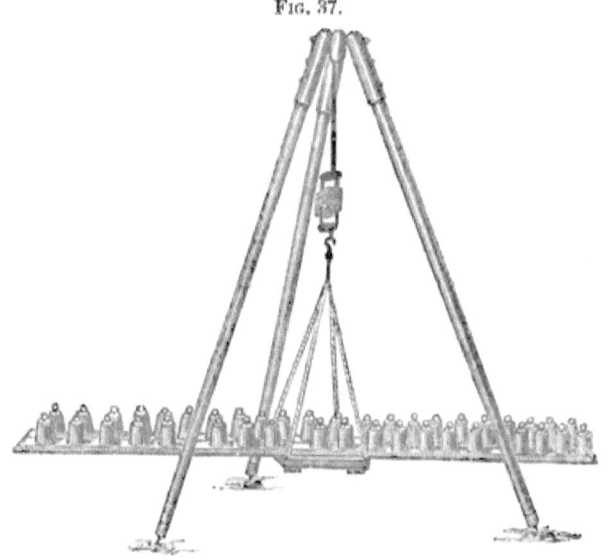

Fig. 37.

or rupture tests. The scale-beam form of machine was the common and usual mode of testing in this country until 1858, when the steelyard machine was first used by the engineers of the Metropolitan Board of Works. The scale-board tester is inaccurate, owing to the inseparable vibration of the platform, as each succeeding weight which was put on swayed it to and fro, and produced results of doubtful and varying accuracy.

Adie's steelyard machine, first used by Mr. Grant, and through the agency of which all his recorded tests were performed, is shown in Fig. 38. This machine is simple in character, and generally indicates with approximate accuracy the tensile value of materials submitted to its scrutiny. There is, however, at the period of examination, some degree of vibration during the progress of the moving weight when traversing the beam. The machine is adjusted, when placing the briquettes with the clips in which they are enclosed, by a hand-screw underneath the table, and this being accurately

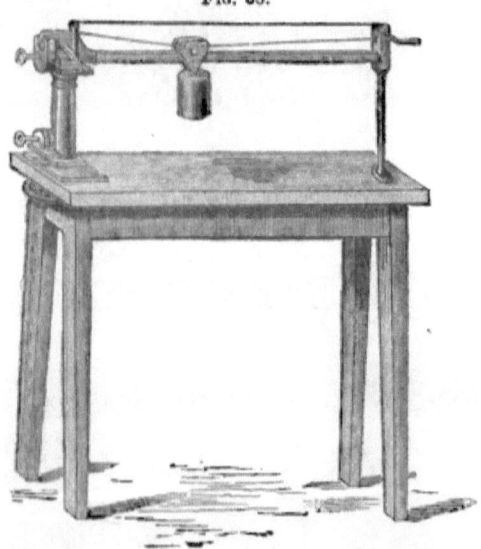

Fig. 38.

accomplished, the small handle at the other end of the machine, on being turned, moves by a catgut line the depended weight, which, when the fracture occurs, indicates the breaking weight on the graduated steelyard or beam. This machine has attained a position from the fact of its having been that used by the engineers of the Metropolitan Board, but it cannot be accepted as a perfect one by the accurate and technical examiner.

Pallant's testing machine, as shown by Fig. 39, is similar in character to that of Adie's, and is now used by the Board of Works. It has attained a high reputation amongst contractors and engineers from its stability and freedom from derangement. These machines have been supplied to the Government engineers and contractors, and used by them on many important works at home and abroad.

FIG. 39.

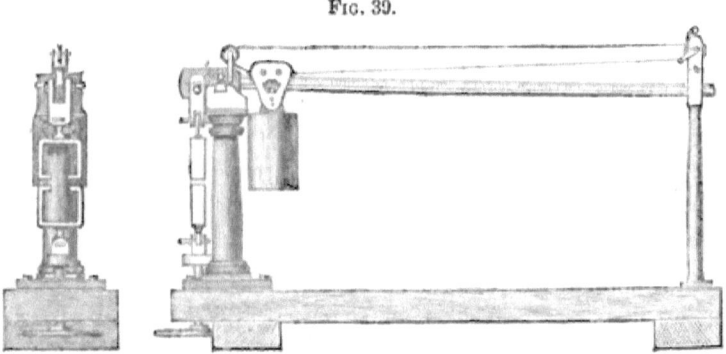

Another testing machine, of a simple and inexpensive character, which has obtained some position amongst home and foreign engineers, is Michele's, shown by Fig. 40. After the adjustment of the briquettes, the rupture is produced by turning the handle which moves the rack and pinion, thus exerting a strain on the briquette clips, raising thereby the circular lever weights. The reading of the tensile breaking weight is recorded by the loose index point on the graduated arc. There is considerable strain in practice in the use of this machine, and consequent vibration during the time of testing. It is, however, for ordinary purposes, a useful and handy machine, where extreme accuracy is not desired or necessary.

Fig. 40.

We will here call attention to a highly ingenious and thoroughly scientific testing machine, invented by Professor Thurston, of America,* which fully, as far as metals are concerned, recognizes and measures every known strain, and most searchingly proves their respective values. Its testing powers are almost unlimited, and indicate by an autographic registry the strength, elasticity, limit of elasticity, ductility, homogeneousness, and resilience of the submitted sample.

In speaking at a recent discussion on Portland cement, Professor Reynolds, of Manchester, the President of the meeting, said, "There was one point with regard to the strength and testing of cement, which appeared to open out a very considerable field for research. The question was as to whether it should be a test of compression, or a test of tension. There was

* Sole makers in England are Messrs. W. H. Bailey and Co., Salford, Manchester.

little doubt that for general purposes compression is the force the structures for which cement is used are called upon to bear. But it is questionable whether the masonry gives way from the crushing of the cement. If the stone is strong, the cement might be called upon to hold the wall together and prevent it from bursting, and so the strength called for might be one of tension rather than compression. It would seem that the tensile strength is the best of the two, because for compression there is a large margin of strength, whatever may be the condition of the material. In tension the whole result is shown. In America they tested cement by twisting, and in this way they got over what is the difficulty with all

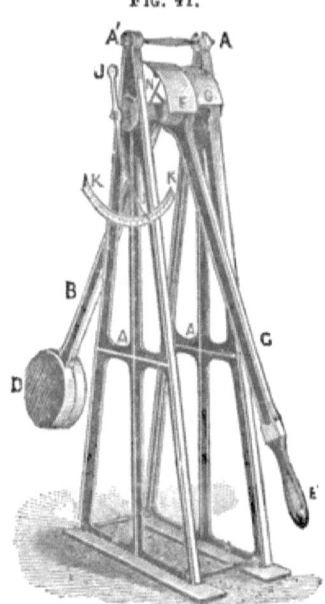

Fig. 41.

tensile strains, and that was the impossibility of bringing the force to bear fairly on the section. By merely varying the line of force one-sixteenth of an inch, it would reduce the strength something like a quarter. This method was introduced into America by Mr. Thurston, and seemed to give very satisfactory results. The character of the force seemed to be similar to that which masonry is called upon to bear. In Liverpool they had a system of testing cement in compression by means of the anchor and chain testing machinery, and in this way found somewhat striking results, not so much with regard to quality of cement as to the way in which the strength was affected by the mixture that was put in. He thought it was a mistake to put in too much cement. It was sufficient to fill the crevices between the stone and sand. By using small stones with larger ones they might fill up the cavities, leaving little room for the cement."

Fig. 41 represents Professor Thurston's machine. The author is not acquainted with any tests of cement obtained through its agency, but hopes at some early period to institute a series of cement and mortar tests by it, and as indicated by Professor Reynolds.

A great obstacle to the practice of testing for ordinary purposes is its costliness, and the machinery required for the simplest tests cannot be commanded by many to whom testing would not only be instructive, but in the long run economical. Danger would be foreseen, and disaster and loss thereby avoided. In the desire to realize the desideratum of a reasonable priced machine, many attempts have been made, but the inducements held out have not yet succeeded in awakening the required interest amongst architects and builders. The apathetic indifference one might almost term it, which induces this class of constructors to trust to the cement maker's assertion that the cement was tested by the "Metropolitan Board of Works Test," or "Government Test," or some other equally apocryphal agency, is indeed humiliating. In Germany, where the cement question receives its well-merited attention, the want of a cheap and reliable tester has now become a matter of importance. To meet this exigent and growing want, a comparatively cheap tensile and compression machine is offered by Messrs Fruhling, Michaelis, and Co., of Berlin, drawings of which are shown by Figs. 42, 43, and 44. These machines do not to our notions of excellent

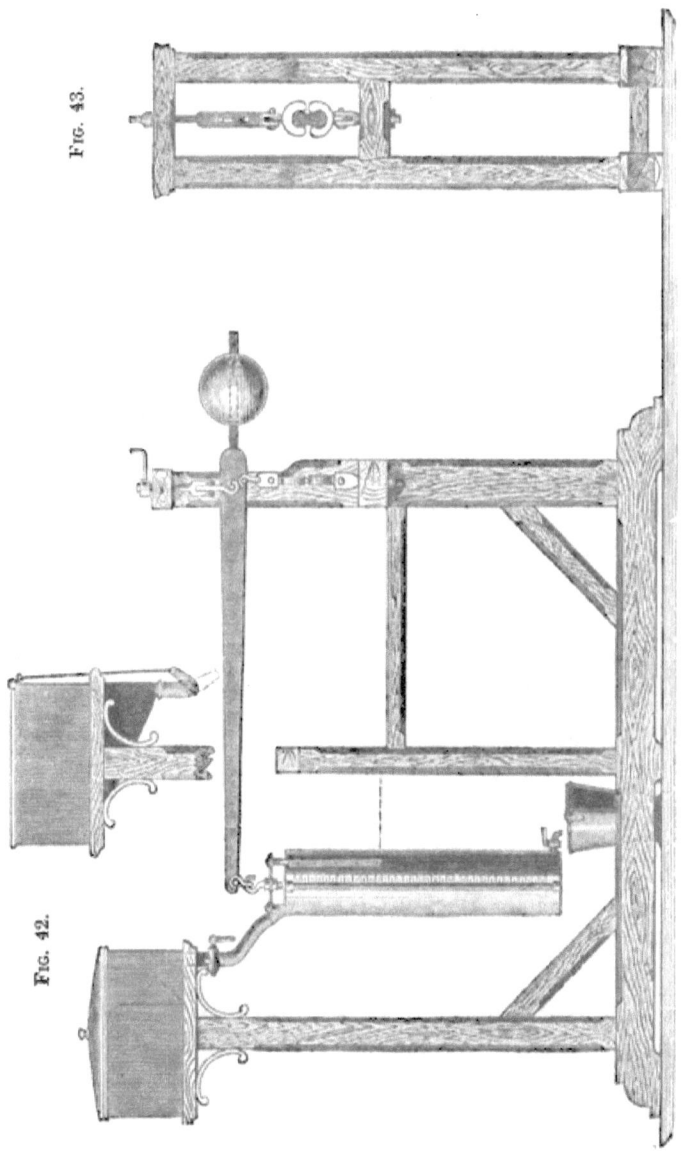

Fig. 43.

Fig. 42.

manufacture appear first class in character, but their cost will go far to commend their adoption, where only approximate accuracy is desired.

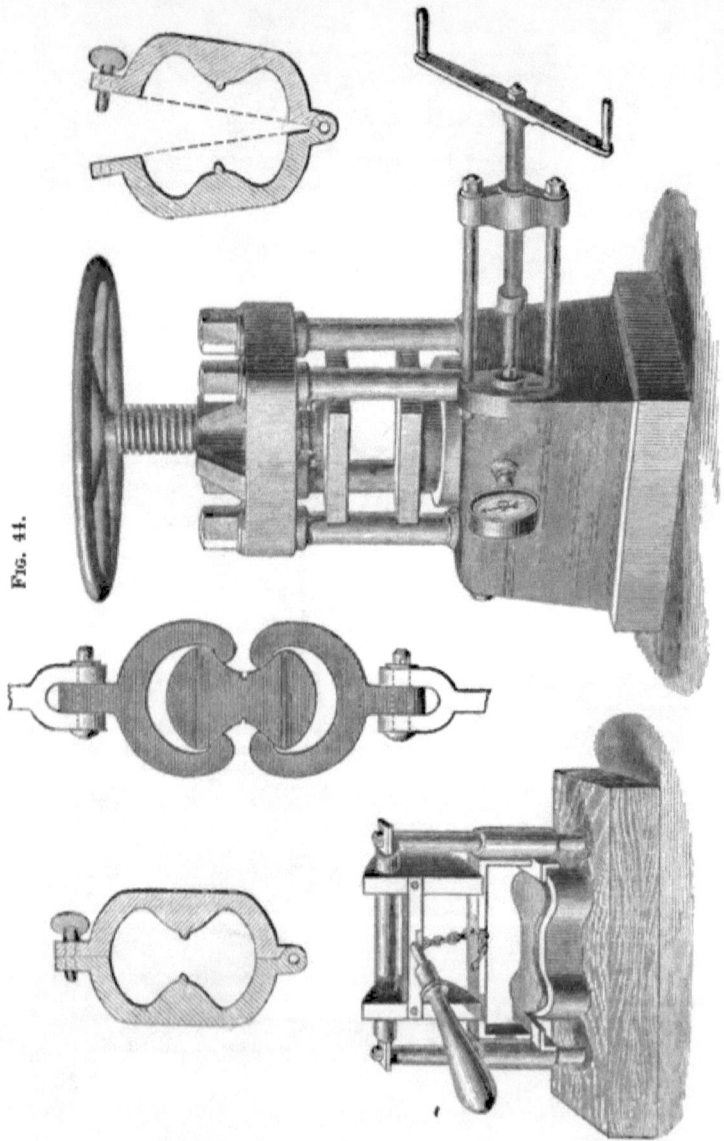

Fig. 44.

The next machine on which we will observe, is that manufactured by Messrs. W. H. Bailey and Co., of Salford, who have had much experience in the make of testing machines of every description. Their appearance and workmanship are more agreeable to our ideas of this class of machinery. Figs. 45, 46, tensile testing machine, and hydraulic compression apparatus. The author has made some tests through the agency of these machines, and found

Fig. 45.

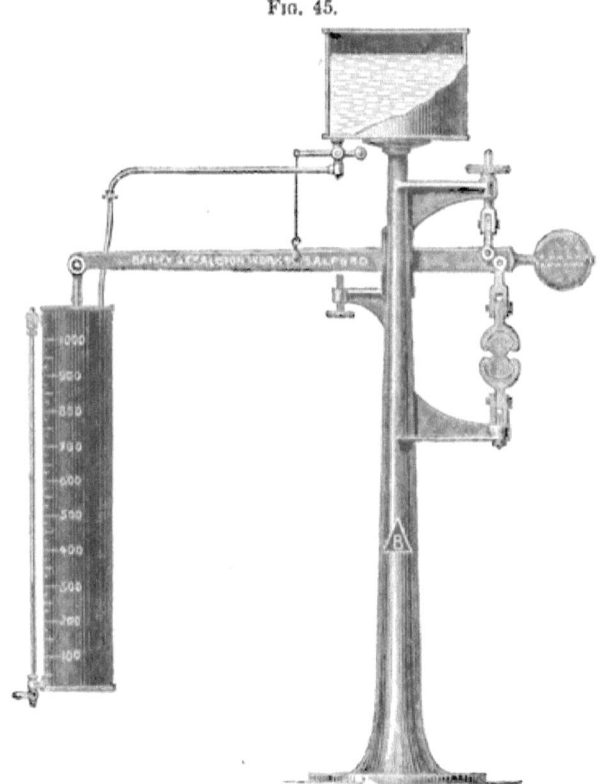

them reliable and accurate. It appears to us that this kind of tester in price and simplicity descends low enough to avoid on the one hand excessive cost, and on the other secures the necessary accuracy without which no machine, however inexpensive, should be tolerated. There can be no question of its extreme delicacy, and absence of vibration, while at the same time it has the advantage of being self-recording.

FIG. 46.

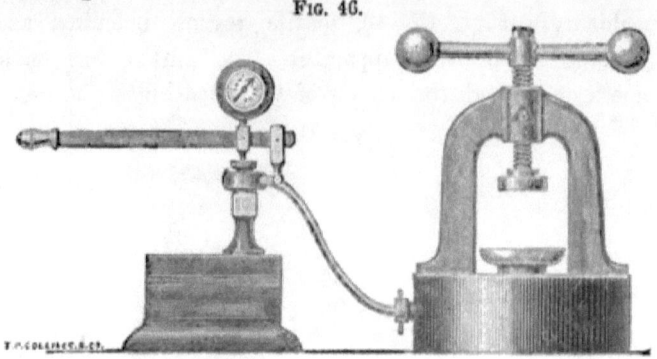

In 1870 the author used at the Rugby Cement Works a testing machine of considerable merit, which he had made at Stettin. It is shown in Fig. 47 arranged for testing briquettes under tensile strain. It was also competent to test the compression value of inch cubes by an alteration of the levers and weights, but we will confine our description to the machine as delineated in the sketch. The proportion of the lever arms is five to one, but by their alteration it can be increased to ten to one; such a change, however, being only necessary when prisms of one inch cube require to be tested for compressive value.

The briquette is inserted into the clips K K, and their adjustment regulated by the screw I, which is fastened to K by a ball-and-socket joint, so as to prevent lateral deviation of the lever while the scale-board is being loaded with weights. The lower clip is connected with the fork L by a hinge, the fork itself being fastened in a steel groove in the lever A by a steel edge. The distance of this groove from the axis is one-fifth of the whole length of the lever. The lever is supported by the screw S. When the briquette is adjusted in the clips, the upper clip is tightened by the nut M, and the lever is then liberated by removal of the screw S.

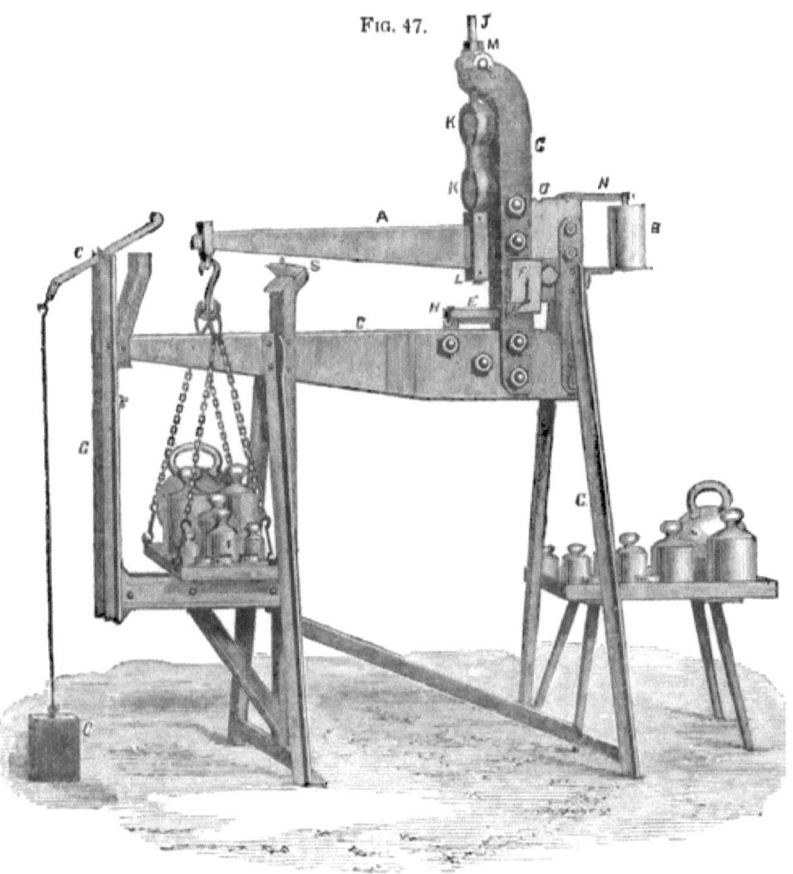

Fig. 47.

It now acts upon the briquette with its own weight, and that of the scale-board. In such cases, as where a test of neat cement is required in a few hours, or one of mortar after a few days, such briquettes would not in all probability bear the weight of the lever and scale-board. The lever

C is then used, and can be so arranged as to neutralize wholly or partially the weight of lever A and scale-board. The scale-board is now loaded with weights till fracture ensues. The form of mould used was rounded, as shown in sketch.

This testing machine is made in a most careful manner, and its appearance and character reflect much credit on the German workmen by whom it was turned out.

The prisms to be tested are placed on the plate E, resting on edges, and the clips K K and fork L are removed. The edges at H on which the plate rests are only represented on the sketch at one point, but the other is exactly 0·13075 metre apart. The lever is now arranged, and is prevented from moving by the weight B acting upon the lever N; a steel edge is then placed in the groove of the lever A exactly vertical and parallel to the sample.

The mode of testing thus hurriedly described is somewhat tedious, and requires much nicety and care in its performance. When great accuracy is desired, and the double test wanted, this machine is one well suited for use in a manufactory, where a special workman should be entrusted with its management. It was so used at Rugby by the author, and it assisted him largely in the necessary experiments when he undertook the task of releasing these works from their ignorant and prejudiced enthralment.

In the beginning of the present year, a general meeting of manufacturers of bricks and cement, and others interested in these materials, was held at Berlin. Among other decisions arrived at by this body, it was resolved, for the purposes of reducing the inconvenience of a variety of standards and conditions, to institute one uniform system directed by suitable rules for using cement. The meeting was almost unanimous in agreeing to adopt the following tests.

1st. The weight of the casks and sacks in which Portland

cement is brought into the market shall be uniform; the manufacturers are only to pack cement in normal casks of 180 kilos. gross, 170 kilos. net weight, half-cask of 90 kilos. gross, 83 kilos. net, and sacks of 60 kilos. gross.

For leakage and differences in the weight of each cask 2 per cent. may be allowed.

The casks and sacks must bear the name of the manufacturing firm, and the amount of gross weight.

2nd. According to the use for which it is intended, Portland cement should be ordered slow or quick setting. For most purposes slow-setting cement can be used, and is to be preferred, because more safe and easy to work, and superior in strength.

A cement may be termed slow, if it requires half an hour or more to set.

3rd. Portland cement must be constant in volume. The decisive test in this respect is to place a thin cake of cement, made upon glass or a tile, in water, observing, after a considerable lapse of time, whether it becomes contorted or cracks show themselves at the edges.

4th. Portland cement must be ground so finely that a sample of it will not leave more than 25 per cent. residue on a sieve *of* 900 *meshes per square centimetre.*

5th. The cementitious strength (tenacity) of Portland cement is to be tested by means of a mixture of cement and sand. The tensile strength is to be tested, according to a uniform method, by means of the briquettes of the same shape and section and the same apparatus. The tests are to be made with briquettes of 5 *square centimetres* section.

These briquettes are to be made in the moulds and apparatus of Dr. M. Leger and Dr. Jul. Aron, of Berlin.

6th. Good Portland cement must, when mixed with three parts by weight of clean sharp sand to one part by weight of cement, and hardened for one day out of, and twenty-seven days in, water, have a minimum strength of 8 kilos. per square centimetre.

The standard sand for this test must be of a certain size of grain, and is procured by sifting sand as it occurs in nature through a sieve of 60 meshes to the square centimetre. The coarser particles are thus excluded, and the sand prepared in this manner is again sifted with a sieve of 120 meshes per square centimetre, to free it from the finest particles.

The briquettes must be tested immediately they are removed from the water.

In the case of quick-setting cements, the tensile strength of 8 kilos. per square centimetre in twenty-eight days cannot be expected. The apparatus used in furtherance of these rules is shown in Fig. 48 and a description of the mode by which it is used is as follows.

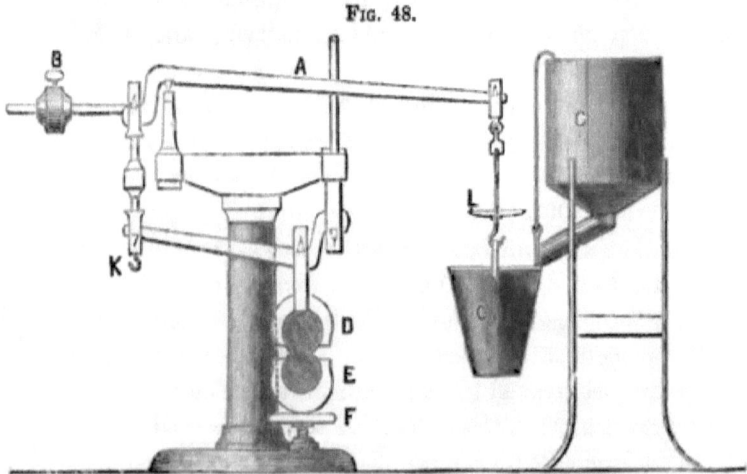

Fig. 48.

The apparatus must be put together as indicated by the above sketch. The lever A, by means of the counterweight B, is brought into such a position that its three knife-edges form one horizontal line. This position is indicated by a mark on the machine. The small bucket C is then hung upon the hook attached to the scale-board L, and the briquette is slid into D and E. In doing this, care must be taken that the four clip ends which are opposite each other are parallel, so that the tension of the briquette is uniform. By means of the wheel F the briquette can be adjusted so that the lever A again attains its former position. The receptacle G is fitted with shot and placed by the side of the bucket C, so that the shot may fall into the latter. For this purpose the slide T is raised by means of the brass wire so that the shot falls quietly, but uninterruptedly, into the pail. The toothed part of the wire is hung upon the knife-edge immediately below the edge of the receptacle. The instant fracture ensues, the supply of shot is cut off by pressing the bent wire outwards, thus removing it from the knife-edge. The weight of the bucket and shot has caused the fracture of the briquette. The weight of this is determined by means of the apparatus itself. The bucket is hung upon the short arm of the upper decimal lever by the hook K, and weighed by weights put upon L. This being a decimal weighing machine, if the number of grammes upon L be a, then the weight of bucket and shot will be $10 \times a$. This latter weight multiplied by 50 gives the weight which caused the fracture, the proportion of the upper lever arms being 1 : 10, and of the lower 1 : 5. The weight which caused fracture is therefore $50 \times 10a$; the section of the briquette being 5 square centimetres at the point of fracture, the breaking weight per square centimetre is $\frac{50 \times 10 \times a}{5} = 100 \times a$ in grammes, or $\frac{1}{10} \times a$ per 1 square centimetre in kilogrammes. If, for instance, 105 grammes were upon the scale-board, the breaking weight would be 10·5 kilogrammes per square centimetre.

We may mention that a change has been made in the moulds for making the briquettes, as compared with those formerly in use. They were previously constructed with hinges and screws, but in order to make the briquettes more uniform the flanges have been furnished with two pegs and a clip to press the halves together. The latter can be easily removed when the mass in the mould is sufficiently set. The halves can be separated from the briquette without difficulty, especially if the inner surfaces are previously slightly oiled.

There is an important consideration which should not be overlooked in any of the testing machines, and that is the absolute necessity of applying the forces of strain in a *direct line*. The most trifling departure from this, regardless of the form of mould, will have a very injurious influence on the result of the test. This error is more likely to arise in the case of simple beam levers, which require considerable depression to cause rupture of the briquette.

CHAPTER XIX.

CEMENT TESTING.

It may very fairly be asked why should consumers of Portland cement be obliged to resort to such trouble in proving this material and protecting themselves against its dangers? This duty surely, in common fairness, belongs to the producer, and after being paid for the article the damage caused by its imperfections should be borne by him. If the cement is imperfectly made, and unsuitable for the purpose for which it was sold and bought, surely the party permitting its circulation is alone culpable, for in more than one of the stages of its manufacture he could have challenged its quality, and so prevented its doing mischief.

There is much difficulty surrounding the question of testing, from the unavoidable irregularity and variableness of the different cements. It would not be simplified, even if it were possible to obtain all the cement from one manufactory. Not only do the raw materials fluctuate in chemical and mechanical value, but the products obtained from their intelligent combination are equally unreliable. In proceeding a step farther, we again encounter differences in value by the same manipulator, and from kindred parcels of cement, when the results of the testing machine are recorded. Under such circumstances obstacles of an almost insurmountable character require to be overcome before accuracy can be secured. The realization of unimpeachable tests by the institution of rigid preliminary technical examination would be too tedious and expensive for the consumer to perform.

The first safeguard should be a thoroughly chemical examination by analysis of the cement after exposure for a reasonable time to the influence of the atmosphere. For the purposes of such scrutiny a standard analysis would be required, derived from the most trustworthy sources, and the allowed divergence from which should be limited and prescribed within reasonable bounds. If such a preliminary challenge could be so accurately performed as to secure a perfect Portland cement, all other succeeding examinations would be rendered unnecessary and superfluous. Such a duty would necessarily require to be performed by one versed in chemical science, and obtainable from an outside source, for on public works or in engineers' or architects' offices such knowledge is at present but rarely found. Until, therefore, chemical knowledge becomes one of the necessities of an engineer's education he must look elsewhere for that assistance which becomes every day more pressing and necessary. Where then are we to look for help in this difficulty? If the manufacturer would guarantee his cement the problem would admit of easy solution, but until this not improbable millennium arrives we must find through some specially trained assistance the needed protection. Secure the chemical accuracy of the cement, and your future path in its use will be an easy and comfortable one, for the succeeding treatment, after it has left the kiln, is one entirely mechanical in its character. Fineness of the powder is readily tested by the agency of a properly made sieve, requiring but a small amount of intelligence for its performance. Should such an ideal Paradise for the engineer and those under his authority ever be realized, the use of Portland cement would reach a point which, under existing circumstances, is impossible, for all distrust would then be removed.

It is only during the last few years that any serious

attention has been given by English consumers to the important subject of testing. The large and interesting works entrusted to the Metropolitan Board of Works offered a favourable opportunity to their engineers for introducing Portland cement in the new drainage and other works. Their experience under the old Commission of Sewers had not afforded them the opportunity of displaying any advanced engineering knowledge in the construction and repairs of the common sewers over which they had control, and, in fact, up to that time, Roman cement was, with but rare exceptions, the only cement they used. This was probably due to two causes, the one being doubtless the distrust attached to the use of Portland cement from its variable and uncertain quality, and the other because the engineers of the old Board had not then the necessary experience or technical knowledge to warrant their adopting so suspicious a material.

For many years the Portland cement manufacturers of England had supplied the French and other foreign Governments with large quantities of cement, and in doing so had to pass it through a rigid system of testing. The onerous conditions attached to the contracts for these supplies necessitated the most careful preliminary tests before the shipment of the cement from England, so as to avoid the danger and risk of its rejection at the port of discharge, which was generally a Government dockyard. It was while engaged in the supply of cement under one of these contracts that the author first became conscious of the importance of testing, and was in a position in 1858, when first consulted by Mr. Grant, on behalf of himself and Mr. Bazalgette, to advise them as to its form and character. Hitherto the briquettes, as directed by the French system, were first moulded in a solid form, and at a prescribed interval of time cut at the sides to receive the clips, and thus eventually submitted to the tensile strain. This cutting involved a waste of time and money, which the author obviated by devising the press (as shown in Fig. 49) now generally used in moulding the briquettes of any form, and thereby preventing the fractures which, by the French method, during the process of cutting, was unavoidable.

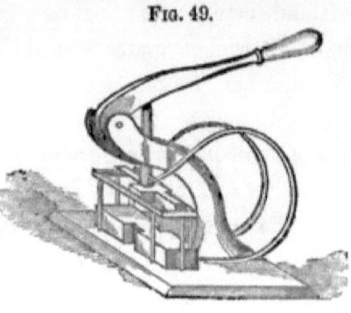

Fig. 49.

Reasonable objection is now made to the form of the existing compound test, and as the author is responsible for its original adoption by the Metropolitan Board of Works, he will avail himself of this opportunity of explaining why it was prescribed.

During the early struggles of the Portland cement makers, who, it must be remembered, were entirely ignorant of the chemistry of their trade, it was found (long after Pasley's rejection of the clinkered mass from the kiln) that the strongest cement was obtained from the heaviest products, and as its manufacture was more costly, it became imperceptibly the practice to charge cement at so much per bushel, according to weight. The schedules of prices in all the contracts for work under the Metropolitan Board had an item for Portland cement at so much per bushel of certain specified weight. Indeed, Roman cement was always, and is still, generally sold by the bushel, and that measure, as the best known one of capacity, became an element in the test of what is now the fashion to designate "the Metropolitan Board of Works Test." The mode again by which the bushel measure was to be filled was never meant to be a fanciful one, or to be surrounded with conditions beyond the capacity of an ordinary workman.

The bushel was to be filled with a shovel from the heap, and it was to be a stroked, not a heaped bushel. The measure was not to be beaten or touched while being filled. One would have thought a measure so well known, and universally used in the estimation of grain and other produce, would have been free from challenge, and possibly might, if those seeking for extra refinement of testing had allowed it to remain in its original position. Its continuance need no longer be pressed if a better and more reliable one can be found. This explanation reminds us of the laboured ingenuity of Mr. Bramwell in proving to the members of the Institution of Civil Engineers (during the cement discussion 1865-66) that a bushel measure of cement could be filled in sundry ways, and with varying and therefore unsatisfactory results. Why no one ever contested the point, and it was adopted under the best advice then available; and you might as well have blamed George Stephenson for adopting the ordinary track of carts and wagons as his gauge for railways, or his first-class carriage after the stage-coach model, although both of these adopted models might have been challenged by any captious critic, who could readily have contended that the coach track was not 4 foot $8\frac{1}{2}$ inches, for the wheels of the cart had an oscillating play on the axles, and the carriage was larger and more comfortable than the coach.

There were other reasons, however, as I have elsewhere fully explained, besides that of an acknowledged and universally recorded measure of capacity, for its adoption. In past times, and probably still in the present, a considerable amount of adulteration was practised by the addition of various slags of high specific gravity to the cement, such addition enabling the mixture to pass the weight test, which, in many cases, is the only one capable of application. Under such circumstances, a measure of capacity test would have been of little value if it had not been succeeded by a tensile one, and again, even the adoption of that twofold test was dangerous, unless completed by the hydraulic one. When adulteration of the kind we have referred to is practised, the best advantage to the manufacturer is obtained by an admixture of light or under-burnt cement, which frequently in that condition contains a large percentage of free lime, that the water test would assuredly detect. Another practice is also resorted to of mixing the contents of an imperfectly burnt kiln with an over-burnt or highly limed one.

The bushel element of the test may now, if found inconvenient, be eliminated, at least from the constructive or consumers' point of view, although some manufacturers may still regard its retention as a measure for ascertaining the comparative costs of their products. Yet as a comparative aid in experiments and for information in their conduct, some test of capacity must still be retained; but it may be of a much more portable character and with greater pretensions to scientific or technical precision. A test of specific gravity is suggested by Mr. Mann to meet the difficulty, and we extract from that gentleman's paper on the "Testing of Portland Cement" the following description of a gravimeter devised by him for that purpose.

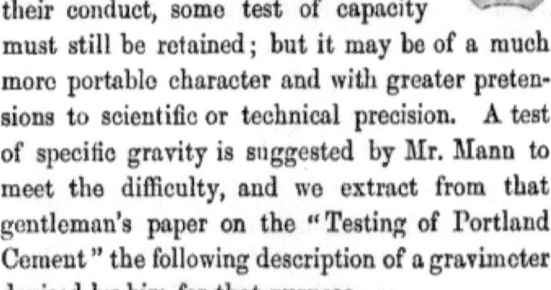

Fig. 50.   Fig. 51.

"It consists of a small glass vessel, holding, when filled to a mark (A, Fig. 50) on the neck, a given quantity of liquid, and of a glass pipette furnished with a graduated stem and stop-cock, and containing, when filled to a mark (B, Fig. 51) on its upper extremity, a volume of liquid equal to that held by the first-mentioned vessel, *minus* the quantity displaced by 1000 grains of the densest substance intended to be examined."

"In using the gravimeter, the pipette is filled to the mark (B) with paraffin, turpentine, spirits of wine, or any other liquid which does not act on the cement, preferably paraffin; 1000 grains of the cement are then introduced into the smaller vessel, which is placed under the pipette and filled to the mark (A). Before this is quite completed, the vessel may be corked, and the contents shaken to remove any small air-bubbles that may be entangled in the cement. The height of the column of liquid remaining in the pipette determines the specific gravity, which can be at once read off on the graduated stem. It is manifest that the denser the substance operated upon, the less liquid will be displaced in the smaller vessels, and therefore the less will remain in the pipette, and *vice versâ*. In reading the accompanying gravimeter, the second place of decimals is estimated. Any greater degree of delicacy may be obtained either by diminishing the diameter of the stem, or by reducing the range.

"The specific gravity of any solid substance coming within the range of the instrument can, of course, be taken in the same manner. The advantages claimed for this gravimeter are, that neither the density nor the temperature of the liquid used need be taken into account; one weighing is sufficient, and all arithmetical calculations are dispensed with; it is also inexpensive and requires little skill in manipulation.

"In order to test the accuracy of the instrument, a small piece of granite was reduced to powder, and its specific gravity taken by the gravimeter; the specific gravity of an unpulverized piece was then ascertained by the ordinary method. Similar experiments were also made by a piece of limestone. The results were:

Granite	Specific gravity by gravimeter	2·62
	Ditto, by the ordinary method	2·63
Limestone	Specific gravity by gravimeter	2·70
	Ditto, by the ordinary method	2·71

"Some of the results obtained in comparing the weights of cement of various manufacturers, as usually taken, with the actual density or specific gravity, are shown in the following table.

Weight of 1 Cubic Foot of Cement as ordinarily obtained.	Specific Gravity.
lbs.	
75·50	2·91
81·50	2·80
83·25	2·96
84·25	3·03
85·00	2·93
85·00	2·96
85·50	2·82
87·50	2·91
89·00	2·96 "

The quality of the cement used in these experiments must have been very light, for, taking the average of the nine samples, we only get 84 lbs. per cubic foot, or the weight per bushel (capacity being 1·28 cubic feet) 107½ lbs., while the lightest, 75·50 lbs. per cubic foot, gives 96½ lbs., and the heaviest, 89 lbs. per cubic foot, gives 114 lbs. per bushel, low weights compared with those obtained by Dr. Finreck, as described at page 252 of "London and Stettin Cements," the average weight being 135½ lbs. per bushel.

"The weight therefore, as ordinarily found, bears no relation to the density. Taking the whole number of experiments (about fifty), the specific gravity varied from 2·77 to 3·03, the average being 2·91, showing that Portland cement is heavier than ordinary building stone. The specific gravity of a number of specimens of fine-sifted cement gave an average of 2·9, that of the coarse particles of the same cements being 2·93. The density of a specimen of unground clinker was 2·55; the clinker contained numerous small air-holes, which accounts for the specific gravity being less than that of ground cement. Some samples of gauged cement, the age of which was three months, had an average density of 2·21, the extremes being 2·07, and 2·45."

This simple and cheap instrument might be profitably adopted in such experiments as the ascertainment of the weights of cements in conjunction with the other elements of tensile and hydraulic value.

The air-holes referred to in the specimens of clinker are the unavoidable result of calcination. Such a vesicular appearance is common to the generality of basaltic rocks, except that the excess of lime in the cement secures a more ready vitrifaction of the artificially prepared compound as well as assists in its eventual capacity of disintegration.

It may be observed that the basalts, as they approach in chemical value to the Portland cement analysis, sympathize also with it in its liability to atmospheric degradation.

Since the above quotation from Mr. Mann's paper was written, the Council of the Institution of Civil Engineers have awarded that gentleman honours for his contribution to the science of Portland cement. Although hailing, with much satisfaction, any aid to this subject, however insignificant, we feel it necessary, in fulfilling the task we have undertaken, to point out any criticisms, from whatever source, which challenge or dispute the accuracy of those quotations obtained from outside sources. In furtherance, therefore, of this duty, we give a translation of a recent article in the 'Thonindustrie Zeitung,' on Mr. Mann's gravimeter. The article is headed "On the Specific Gravity of Portland Cement."—"Experiments were made by us as to the specific gravity of different cements with a Mann's apparatus, and to our surprise we arrived at the results below.

"Although the apparatus is but little applicable for accurate scientific experiment, yet it is very serviceable for comparative experiments which require dispatch.

"We first took pulverized quartz and determined its specific gravity by means of distilled water. Fifty grammes were used for this experiment. Three trials gave identical results; the specific gravity being 2·59. As we thought that air might be retained by the powder, we inclined the flask containing the powdered substance covered with liquid in such a manner that any air-bubbles rising would collect in one spot. The flask remained in this position for twenty-four hours without any air collecting, even when the flask was carefully shaken. The volume of the few small bubbles that appeared in one or two cases was so small that the error in reading off the results would be greater. If the cement is used in the form of a sandy powder, there is no fear of inaccuracy from this source. In estimating the specific gravity of cement, petroleum was used instead of water, as the latter decomposes cement. The air escaped without difficulty when petroleum was used. The following specific gravities were obtained:

Cement from	Groschowitz	..	..	..	3·03
	Wildau	..	..	..	3·01
„	Höxter	..	..	..	3·08
„	Stern	..	..	..	3·08
„	Lüneburg	..	..	..	3·03
„	Beckurn	..	..	..	3·04
„	Offenbach	..	..	..	2·99

The differences in the figures are so slight that they may almost be considered as identical, especially if we take into account that slight experimental errors are inevitable. It is evident that these figures afford no data for the valuation of cement.

"We thought that slightly burnt cement, or in other words, an hydraulic lime of the same composition as Portland cement, would have a different specific gravity.

"Dr. Heintzel had the kindness to send us lightly burnt Lüneburg cement. It was burnt till all carbonic acid had been expelled; but it was still quite porous.

"Two experiments, which coincided in every respect, showed its specific gravity to be $3·04$, the same as well-burnt Portland cement.

"It follows that Portland cement acquires its high specific gravity before it is completely burnt, before, therefore, the clinkering or incipient fusion has reduced it to a dense slag-like product. We may conclude that that which we call Portland cement at a later stage is already produced before clinkering takes place. The object of burning it still more is only to bring the particles closer, to destroy the pores. It is evident that this destruction of the pores must be of the greatest importance as regards the action of water on the cement, as well as the manner of setting. The nearer the particles of cement are brought, the less is the exposed surface, the slower and more regular, consequently, the deposition of the cementitious products of decomposition.

"From the results obtained, so far as a general conclusion may be drawn from them, we must infer that the specific gravity neither indicates the quality of Portland cement, nor whether one and the same cement is lightly or heavily burnt. It may appear strange that cements made from such very different raw materials as the above, namely, fresh-water limestones, chalk, or compact limestone, should show no difference in specific gravity; but we must bear in mind that the ultimate composition of normal Portland cement only varies within very narrow limits. There is, therefore, no reason for any considerable variation in the specific gravity. If variations occur they must be attributed to the greater or less proportion of iron, magnesia, &c. But as these substances occur in very small quantities as compared with the lime and silica, which *form the chief mass of the cement in nearly constant proportions*, it follows that the variations cannot be considerable. If Mann found differences from $2·77$ to $3·03$, he may have experimented on

Roman cement as well, the chemical composition of which differs materially from that of Portland cement."

If these experiments should be hereafter confirmed, the proposed new test of specific gravity and its instrument of measurement will not yet supplant the "old" bushel gauge. These experiments, we have reason to believe, were made by expert chemists, or, at all events, by their direction, and there is, we fear, little likelihood of their being disputed.

It shows only the danger and inconvenience of departing from well-established forms of tests, and the apparatus by whose agency they had been performed.

Liebig recommends as a ready means of testing the specific gravity of sand the following method:

"Take a glass tube, carefully graduated, to denote its capacity in cubic inches, every cubic inch being again divided into one hundred parts. Fill this tube to half its capacity with water, and project into it a weighed amount of sand, or of the substance under examination. The water will rise in the tube. The difference in the level of the water thus produced indicates exactly the volume of the sand in cubic inches and $\frac{1}{100}$ths of a cubic inch."

Fig. 52.

Of course paraffin would have to be substituted for water, when the specific gravity of cement was to be thus ascertained.

While on the subject of specific gravity, we may as well refer to the now old-fashioned instrument or apparatus called "Nicholson's Portable Balance," as shown by Fig. 52.

The body of this instrument is a hollow cylinder of tinned iron, of which each extremity, a and b, terminates in a cone. From the vertex of the upper cone a small stem of brass, $a\ c$, rises perpendicularly, bearing on its upper ex-

x

tremity a small tin cup, d. From the vertex of the lower cone is suspended a similar cup, e, attached to a cone of lead underneath it as a ballast. Both the cups may be removed when the balance is not in use.

When this instrument is placed in a vessel of water a portion of the cylinder ought to swim above the surface of the water. The tin cup d is then to be loaded with weights till the instrument sinks so far that the surface of the water may exactly coincide with a mark near f on the brass stem. The quantity necessary to make the instrument sink thus far may be marked on the cup, as a given quantity for future use. Suppose this quantity to be 600 grains, which may be called the *balance weight*, it will serve for taking the specific gravity of any substance whose absolute weight is not greater than that of the balance weight.

To ascertain the specific gravity of a mineral, place it alone in the upper cup, and add weights till the mark on the stem coincides with the surface of the water; and suppose this to be 210 grains. Subtract the 210 grains from the balance weight of 600 grains, and the remaining 390 grains is the absolute weight of the mineral in air. Let the mineral be now removed to the lower cup; but as it weighs less in water than in air, the mark on the stem will rise a little above the surface of the water. Additional weights must now be placed in the upper cup till the mark on the stem again coincides with the surface of the water. Suppose this to be 80 grains, which of course will be the weight of a quantity of water equal in bulk to the mineral. We now have the absolute weight of equal bulks of water and the mineral; then say, as $80 : 390 :: 1\cdot000 : 4\cdot875$.

In ascertaining the specific gravity of Portland cement by this instrument, it would be necessary to use paraffin, or else enclose the cement within a waterproof envelope. Through the agency of such an instrument the specific gravity of any mineral can be readily ascertained. Of course in those substances which sensibly absorb much water it will be necessary to determine the amount of absorption by re-weighing in the upper cup, and add the difference thus ascertained to the first amount.

Hitherto, and generally, the practice followed in the preparation of the briquettes is to mould them in the bell-metal moulds, placing them in water the following day, where they continue for six days. The accuracy with which this apparently simple operation is performed influences in a remarkable degree the succeeding results, and may account in a great measure for the eccentricity of the tests obtained by Mr. Grant, and recorded in the published minutes of the Institute of Civil Engineers. Much attention has been directed by thoughtful engineers to the discrepancies which these tables of tests indicate, and the impossibility of establishing any fixed or reliable rules or formulæ through their aid.

In some experiments made with Portland cement from Hanover it was found that the briquettes of neat cement, after having been in water for three months, when tested after being exposed to the air, exhibited the following results.

When broken immediately on being taken out of the water, the rupture was effected by a strain of 43·2 kilos. per square centimetre.

After 1 day	.. .. ..	30·0 kilos. per square cent.	
,, 2 days	.. .. ..	17·1 ,,	,,
,, 4 ,,	.. .. ..	33·0 ,,	,,
,, 5 ,,	.. .. ..	40·2 ,,	,,
,, 6 ,,	.. .. ..	40·8 ,,	,,
,, 7 ,,	.. .. ..	44·4 ,,	,,

Should these experiments be confirmed, another new element of inaccuracy must be added to the already numerous dangers surrounding the testing of cement as generally adopted.

In briquettes made with 1 of the same cement to 1 of sand, no such deterioration was developed, but, on the contrary, they continued to improve in tensile value on exposure to the air.

The length of time and the surrounding climatal influence exert a prejudicial effect on Portland cement when warehoused. In fresh and good Portland cement none of the analyses should indicate the presence of carbonic acid, and probably in a well-made cement it can only exist in an infinitesimal degree. But every cement, even when free lime is entirely absent, attracts carbonic acid on exposure to the atmosphere.

Mr. D. L. Erderienger, in his experiments made some time ago, obtained the following result:

The cement with which he experimented had an analysis of:

Silica	23·7
Alumina and oxide of iron	9·0
Lime	65·0
Carbonic acid	

When this cement was exposed in a warehouse for eight months, in a layer of 3 centimetres deep, it showed the following analysis:

Silica	23·2
Alumina and oxide of iron	8·8
Lime	63·5
Carbonic acid	2·2

At five months old it was:

Silica	23·3
Alumina and oxide of iron	8·8
Lime	63·8
Carbonic acid	1·8

The lime is capable of accurate estimation by chemical analysis, but the other constituents differ too slightly for reliable ascertainment.

When cement absorbs carbonic acid, certain definite changes take place which, when exhibited, influence its behaviour on being mixed with water. Generally speaking, cement improves with limited exposure to the atmosphere, but too much carbonic acid decreases its strength. Fresh cement has the highest specific gravity.

The cement used in the above experiments when fresh, had a specific gravity of $3\cdot20$, and after its absorption of $1\cdot8$ per cent. of carbonic acid, this was reduced to $3\cdot00$, and was still further reduced to $2\cdot96$ when the absorbed carbonic acid amounted to $2\cdot2$ per cent. A celebrated English cement having a specific gravity of $3\cdot09$ when fresh, sunk to $2\cdot85$ when stored in a room for twelve months (the layer of cement being about 1 decimetre); during that time it had absorbed $2\cdot1$ per cent. of carbonic acid.

Freshly burnt cement showed, in by far the greater number of cases, a considerable increase of temperature when mixed with water. In determining the temperature a tin vessel was used, 6 centimetres high, and 4 centimetres wide. In this 60 grammes of cement were stirred by means of a strong iron wire with 20 cubic centimetres of water, and the thermometer sunk to the bottom of the vessel. Care was taken to determine the temperature of the sample previous to mixing. The thermometer must be cylindrical, and should not have a globular end, the glass requiring to be strong.

The 60 grammes cement fill when in a loose state a vessel of 40 cubic centimetres capacity, so that approximately $0\cdot5$ volume of water is added to 1 volume of cement. For the purposes of comparison the same amount of water must always be added to the same weight of cement, as the increase of temperature becomes greater as the amount of water added decreases. In the following table the time is given at which in each case the maximum temperature was reached:

	1 vol. Cement with 0·5 vol. Water.		1 vol. Cement with 0·333 vol. Water.	
1.—	2·5° in	70 minutes	5·0° in	25 minutes.
2.—	6·0° ,,	100 ,,	13·0° ,,	20 ,,
3.—	7·0° ,,	105 ,,	13·5° ,,	25 ,,
4.—	6·8° ,,	14 ,,	9·5° ,,	10 ,,
5.—	11·5° ,,	31 ,,	19·0° ,,	25 ,,
6.—	2·0° ,,	7 ,,	4·5° ,,	5 ,,
7.—	13·3° ,,	18 ,,	17·0° ,,	12 ,,
8.—	9·0° ,,	15 ,,	12·0° ,,	10 ,,
9.—	4·5° ,,	16 ,,	12·6° ,,	120 ,,
10.—	5·0° ,,	5 ,,	7·4° ,,	2 ,,
11.—	6·6° ,,	102 ,,	13·0° ,,	20 ,,
12.—	7·0° ,,	105 ,,	13·5° ,,	25 ,,
13.—	8·0° ,,	90 ,,	11·0° ,,	65 ,,

The amount of temperature is found to increase with the fineness of the cement.

These three series of experiments—viz. 1st, the deterioration of the briquette after being taken from the water; 2nd, the degradation of the cement by lengthened exposure; 3rd, the influence of varying quantities of water with which the cement is mixed—indicate some of the dangers surrounding the accurate estimation of the strength or tenacity of Portland cement. These are, however, within the reach of our control, and are therefore valuable for our guidance. There are enough puzzling inconsistencies in the published results so entirely beyond our management as almost to excuse us in disregarding such information as the above. It will, however, be advisable in all our experiments to examine the conditions which have attracted the observation of these several experimenters.

Although as a general rule the heavier the cement the stronger it is, and Mr. Grant's summarized table, in which he presses this point, is as shown in column 1; Mr. Colson, in column 2 (converted into bushel weights), to show its

CEMENT TESTING.

unreliability; Mr. Mann, in column 3, exhibiting marked inconsistencies.

1.		2.		3.	
Mr. Grant. (Published in 1865.)		Mr. Colson. (Published in 1875.)		Mr. Mann. (Published in 1877.)	
Weight per bushel in lbs.	Breaking Weight of 2·25 sq. in. in lbs.	Weight per bushel in lbs.	Breaking Weight of 2·25 sq. in. in lbs.	Weight per bushel in lbs.	Breaking Weight of 2·25 sq. in. in lbs.
106	473	112	747	102·6	839
107	592	113	688	103·9	794
108	650	114	719	105·2	940
109	647	115	708	106·5	893
110	708	116	729	107·8	897
111	694	117	675	109·0	911
112	687	118	745	110·3	861
113	702	119	706	111·6	1041
114	700	120	718	112·9	897
115	705	121	702	114·1	918
116	768	122	732	..	..
117	718	123	676	..	..
118	644	124	631	..	..
119	788	125	545	..	..
120	732	126	603	..	..
121	706	127	558	..	..
122	717	..	..	..	..
123	674	..	..	..	..
124	820	..	..	..	..
125	816	..	..	..	..
126	657	..	..	..	..
127	865	..	..	..	..
128	917	..	..	..	..
129	920	..	..	..	..
130	914	..	..	..	..

These different results by engineer experimenters for practical purposes are difficult of reconciliation. Mr. Mann's tests are the most satisfactory, but even they fail to establish an accurate dependence on obtaining tensile value from the specific gravity element. He seems, however, to have obtained the best cement according to the existing standard, for with an average weight (taking his 10 tests) of 108·4 lbs. per bushel, he reached an average breaking result of 889 lbs. per 2·25 square inches. Mr. Grant with an average weight per bushel of 118 lbs. obtained 728·5 lbs. tensile strength, on a breaking surface of 2·25 square inches. A more striking difference is, however, apparent if we take Mr. Grant's first 10 tests and compare them with Mr. Mann's.

The average of the first 10 was 110·5 lbs. per bushel, and the breakings 655·8 lbs. per 2·25 square inches. To put the comparison more clearly, we will take the results obtained by these three sets of experiments, and formulate them as under:

Mr. Grant.		Mr. Colson.		Mr. Mann.	
Average Weight per bushel.	Average Breaking Weight on 2·25 section.	Average Weight per bushel.	Average Breaking Weight on 2·25 section.	Average Weight per bushel.	Average Breaking Weight on 2·25 section.
lbs.	lbs.	lbs.	lbs.	lbs.	lbs.
118	728·5	119·8	680	108·4	889

and the first 10 of* each table gives:

| 110·5 | 655·8 | 116·5 | 713·7 | 108·4 | 889 |

After making some allowance for the improvement in the make of cement between the time of the first of these experiments and the last, there still remains a considerable margin of discrepancy for which there is much difficulty in accounting. By a seven days' test it is practically impossible to obtain from the heavier cements the full measure of their maximum strength. There is, however, much satisfaction in the fact that the tests in 1865 are eclipsed by those of 1875, and they again are put in the shade by those of 1877. The first experiments or tests of Mr. Grant were conducted in a somewhat stereotyped manner, and without chemical or other technical guidance, whereas those of Messrs. Colson and Mann appear to have been undertaken and carried out with a full knowledge of the subject, and with the primary object of clearing up the inconsistencies of the 1865 results. Mr. Grant did not pretend that his experiments were of an accurate technical character, but simply of such a kind as

* Mr. Colson's results from the heaviest cement indicate that the full value had not been realized either from the effects of too slow setting or overburning.

were capable of being performed by an ordinary workman or unskilled labourer. Neither Mr. Grant nor the men by whom the tests were made pretended to, nor did they really possess, any reliable technical knowledge of the subject, but merely exercised such care as they considered necessary to ensure that the cement required in that portion of the drainage works over which Mr. Grant was appointed sub-engineer should be free from blemish, and not likely to damage the work in which it was used. The other sub-engineers under the Metropolitan Board of Works had the same tests to perform in their several divisions, but they did not take any steps to place them before the public. These very costly experiments were made at the expense of the metropolitan ratepayers, and the Board should have themselves published what was clearly public property. Instead of this, however, one of their sub-engineers was permitted to place these results in the "Institute of Civil Engineers," by whose rules the outside public are precluded from deriving any benefit except under the privilege of the Council's permission.

The subject of testing cement, and the mode of its performance, is now beginning to attract more serious attention than formerly; and, in consequence, less dependence is placed in the maker's name, which in the past careless and ignorant times had a magic influence, especially when blazoned in a lazy or confiding architect or engineer's specification. The important interests confided to engineers necessitate a conscientious performance of their duty, and more especially in large cities and boroughs, the superintending officer to whom is entrusted the execution of heavy public works, should feel the extent of this responsibility. To show with what interest the question of Portland cement and its uses is now regarded, we will refer to a paper read by Mr. Deacon, Borough Engineer of Liverpool, before the

Liverpool Polytechnic Society of that town, on the 20th November last. It should be stated that when Mr. Deacon became engineer of that town, the use of Portland cement was comparatively limited, and the supply was entrusted to a firm in Liverpool, who were either the agents of a London maker, or else directly purchased from that source. Testing cement at that time was carelessly performed, or entirely omitted, and Mr. Deacon, when referring to this point, says: "The specification at present adopted requires a tensile strength of 800 lbs. per $2\frac{1}{4}''$ breaking surface. The actual strength of the cement obtained during the last year under this specfication has averaged 871 lbs., and since the tests were instituted, the price paid for it by the Corporation has been considerably decreased." Before proceeding to examine Mr. Deacon's proposition for a cheap and ready means of testing, we will shortly allude to his observations on the manufacture of cement, showing that he is not content with a knowledge of the cement, but also understands the processes by which it is obtained. His account of the manufacture in the paper referred to is, on the whole, pretty correct, and we hope he will pardon our putting him straight on one or two manufacturing points. In speaking of the raw material, he says: "There is always a tendency, moreover, to use too large a proportion of the cheaper material, clay; fortunately, however, this defect is readily discovered by gauging a small portion, placing it in water, and then examining it. If too much clay be present, the colour will be a warm brown, as opposed to the cold grey tint of good cement. Such cement sets quickly, but never attains great hardness, and its strength is comparatively low."

The too free use of clay is due to another cause, for it is really much the dearest of the raw materials used; and, by a careless or ignorant maker, adopted to avoid the dangers incidental to an over-limed cement. This practice, before the institution of testing, was almost general, and cement thus made could withstand the water, and a low tensile test, but failed if anything in excess of 100 lbs. per square inch was required from it. The colour, or rather discoloration (for good Portland cement is of a uniform grey colour), is not easily detected, unless when broken after the sample has been immersed some time in water, but an air sample exhibits all the intensity of its spuriousness.

In another paragraph Mr. Deacon says:

"Portland cement clinker, after being ground, always contains more or less lime, which, not having combined chemically with the other ingredients, commences to slake by absorbing moisture from the atmosphere. This slaking process, technically known as purging, must be completed before the cement is used, or it will be greatly weakened and deformed by the swelling of the particles of lime. In order to ascertain if any sample has been properly purged, it is only necessary to make up a disc of cement with the smallest quantity of water, thin out the edges, and place it in water. If on the following day there are no hair-like cracks round the edges, it may be taken for granted that the cement contains no injurious quantity of free lime."

Of course a clinker of such a character as that described is not a Portland cement one, which theoretically should in its conversion during calcination have converted the carbonate of lime into silicates. The presence of free lime thus unconverted is now frequently due to an over-dose of carbonate of lime in the cement mixture to enable it to pass successfully the modern onerous tests. The indications of faultiness in the disc might not be apparent in twenty-four hours, and yet be an inherent source of danger for weeks before its faulty development would appear. The ascertainment by a speedy and simple examination, such as Mr. Deacon suggests, is only of an approximate character, and should not be substituted under even the most hurried circumstances for the tensile compound ordeal, such as the Liverpool Corporation Test, the specification of which is as follows:

"The cement must be of uniform quality, and capable of bearing the following tests to the satisfaction of the engineer:

"1st. Samples of the cement being sifted through a number 50-gauge wire sieve must not leave a residue of more than 10 per cent.

"2nd. Samples of pure cement will be gauged with water, and placed in the brass moulds used by the Corporation within twenty-four hours; the casts thus made will be immersed in still water, in which they will remain until the expiration of the seven days from the date of moulding, when they will be taken out of water, and tested to ascertain their tensile strength, which must not be less than 800 lbs. on the sectional area of $2\frac{1}{4}$ square inches.

"3rd. In ordinary, the cement will be distinguished as quick setting or slow setting. The slow-setting cement, when gauged neat in the moulds, must not become firm in less than three hours. The quick-setting cement must assume a firm condition within half an hour. The test for firmness will be that of resistance to the finger-nail, the test at present adopted in the department.

"The above tests will be applied after each delivery, and should the result be unsatisfactory to the engineer, he may, having given notice to that effect to the contractor, within eleven days of delivery, require the whole of such cement to be removed at the contractor's expense.

"With each delivery the contractor shall send a memorandum of the quantity delivered, and the name of the manufacturer.
"GEORGE F. DEACON,
"*Borough and Water Engineer.*"

The slow and quick setting distinction is a little puzzling. We presume, however, that both equally meet the requirements of Mr. Deacon's test; but as there is no allusion to their respective specific gravities we are unable to account for their common behaviour, unless it is that the light cement is specially prepared for this unusual test.

On the unavoidable difficulty of instituting a test of this quality and exactness on small works, Mr. Deacon suggests what he calls a "simple mode of testing."

"In Fig. 53 is shown an arrangement of testing, which any clerk of works can put together in a few hours at a cost of a few shillings. The block of cement, C, is to be tested by transverse breaking, and the specification requires that

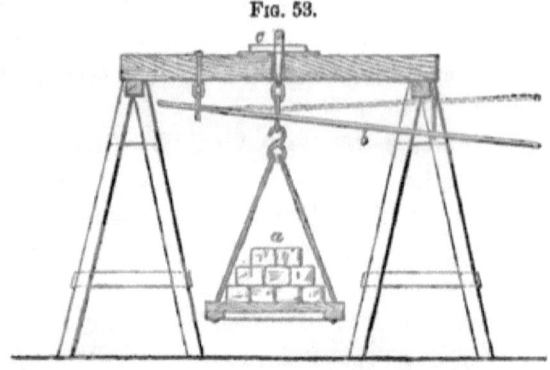

Fig. 53.

it shall not break with a less weight than 150 lbs. applied at the centre. One hundred and fifty pounds of bricks or materials are placed on the board (*a*); it deflects the iron or wooden bar (*b*), but is prevented by that bar from bearing on the shackles hanging from the test block until a man lowers the end (*b*). This he does very slowly, the bar unbends, and the test block gradually takes the load it bore. If the whole load is carried for a few minutes the specific strength has been reached. The dotted lines indicate the position of the bar when at rest.

"The block C used for this test by the author (Mr. Deacon) has a section of 1½ inch square, its length is 10 inches, and the supports are 9½ inches apart.

"The mould may be of wood, and can be put together by any joiner, and the manner in which the block is released from the mould will be readily understood from the sketch.

"Several blocks from each lot of cement should be prepared, and when the moisture has disappeared from the surface they should be placed in water, and tested after seven days.

"In practice the following specification would secure a cement of about the same strength as that employed by the author.

Proposed Specification for Transverse Test.

"Samples of pure cement shall be gauged with water, and pressed into a mould $10'' \times 1\frac{1}{2}'' \times 1\frac{1}{2}''$ The block of cement thus formed shall be placed in water, and after seven days tested by placing it on supports 9½ inches apart, and gradually applying a load of 150 lbs. on its centre.* If more than one out of three blocks are broken within one minute of the application of the load, the engineer (or architect) shall be empowered to reject the whole of the cement from which the samples were taken."

In Mr. Deacon's experience, and from his point of view, slowness of setting is a safe quality in Portland cement, and he says:

"In certain classes of works it is extremely difficult to place mortar or concrete *in situ* immediately it is mixed. Take, for example, the case of a carriage-way foundation 25 feet wide. The excavated materials must be removed from and the concrete materials brought to the face of the

---

* Mr. Deacon regards the transverse strain of 150 lbs. equal to a tensile test of 800 lbs. to the 2¼ square inch.

work by the same route, and if the concrete is to be immediately laid, the mixing apparatus and the men employed in the work must be accommodated on the narrow excavated ground, and must be moved back as the work proceeds."

The great advantage which would arise if the mortar or concrete could be brought to the ground ready mixed must be apparent to all. To ascertain if this was possible without deteriorating the quality of the work, Mr. Deacon, in 1872, prepared twenty-four blocks of concrete of 1 cubic foot each, and allowed different periods to elapse between gauging the material and moulding it. These blocks were tested by crushing, at the material testing machine in Birkenhead, belonging to the Mersey Dock and Harbour Board, with the following results:

"*Results of tests as to the resistance to crushing of Portland cement concrete blocks. Each block 1 foot cube. Proportion of gravel to cement, five to one.*"

INTERVALS IN MINUTES BETWEEN GAUGING CONCRETE AND FILLING MOULDS.

	0	20	30	40	45	50	60	75	80
Resistance to Crushing, in tons, after 28 days	48·1	50·0	50·0	50·0	..	47·15	42·15	..	..
Ditto, after 26 days	43·0	48·0	50·0	47·17	..	47·0	48·15	..	..
Average	45·55	49·0	50·0	48·58	..	47·07	45·15	..	..
Ditto, after 24 days	49·15	..	50·0	..	50·0	..	49·12	43·1	37·1
Ditto, after 21 days	50·0	..	50·0	..	50·0	..	46·0	50·0	50·0
Average	49·57	..	50·0	..	50·0	..	47·56	46·55	43·55

"It is unfortunate that the power of the machine was limited to 50 tons, and that out of twenty-four blocks eleven were therefore unbroken. But the result proved beyond doubt that, on the average, those blocks, the concrete of which had stood from twenty to thirty minutes before being placed in the moulds, were actually stronger than those in which the concrete was immediately moulded. The author is only able to explain this on the assumption, that even in this good and well-purged cement there was some free lime which began to slake on the application of water, thus producing numerous but invisible points of weakness in the blocks first moulded, while in those made with concrete, allowed to stand after gauging, the materials were not moulded until the slaking process had ceased.

"Since the date of these tests about 100,000 yards of concrete have been laid, after being mixed in, and carted from the nearest available yard. The rate of progress of the work has been greatly increased, and the results have been in all respects satisfactory. It is necessary to remark here that, although this is a valuable quality of good Portland cement, it would be most dangerous to take advantage of it without making quite certain that the cement is good and not quick setting."

We would ascribe the increased value of the blocks—made from the concrete kept before moulding—to another cause, which we believe to have been due to the increased compactness of the mass favoured by the partial evaporation of the moisture and its own weight. These two apparently trivial circumstances improved the density of the concrete, and reduced the interstitial space considerably during the interval If the concrete had been carelessly made with an excess or superfluity of water the results of these experiments would have been very different. They were controlled, however, by an engineer who apparently does not consider it *infra dig.* to attend to the smallest minutiæ of his duties. Were there only a few more engineers so disposed we should soon cease to hear complaints about Portland cement and its ascribed mysteries and defects.

The advantage of being able to command a simple and cheap mode of testing, according to the plan proposed by Mr. Deacon, at first sight appears to be a step in the right direction, and meets the complaints of small consumers, who say "that the testing machine is too expensive, and cannot be used on ordinary architectural works, so that those so circumstanced got only the refuse of cement which had been rejected by the large consumers." But only under such conditions would we advise its use, and so long as another more precise and cheap machine of a portable character was not to be had. The present discussion of this important question will show that much advance has been made in the machinery of testing, and less costly machines are now available, besides a better knowledge of what is required and how the test should be performed.

For the purpose of encouraging testing and assisting consumers of cement in Liverpool and its neighbourhood Mr. Deacon has induced the Corporation to allow their testing machinery to be used on the following conditions. He says:

"With regard to the testing at the municipal offices, he wished to explain that the Corporation, at his recommendation, had allowed these tests to be made on payment of ten shillings in each case, with the understanding, however, that the result of the test should not be stated or delivered in writing, though anyone interested might see the test."

The object of presenting written certifications was to prevent dealers or makers of cement from advertising that their cement passed the tests of the Corporation of Liverpool. This practice was once if not still fostered and encouraged by the engineers of the Metropolitan Board of Works, with the too unavoidable and reprehensible result of advertisers proclaiming in loud voice that their cements passed the "Metropolitan Board of Works Test," and could produce in evidence a letter from the tester, or some other equally inconsiderate public officer, as proof of their assertion. Any common-sense person would of course estimate such a certificate at its true value, and fail to be persuaded that a test of a few pounds of cement made by a labourer in the employ of the Metropolitan Board of Works would be a safeguard for all time in the use of cement from whatever factory it issued, or under whichever banner or label it was sold.

In the papers of Messrs. Grant and Colson there is unfortunately too prominent allusion to the makers of the cements, and if we were not satisfied that these gentlemen were actuated solely by an honest desire to forward the subject in a professional sense only, we might feel disposed to think the papers advertising mediums of the makers therein named. Mr. Mann has wisely refrained from this practice, and we trust that all succeeding investigators will follow his laudable example.

The fineness of the cement has much influence in the resulting tests as well as its behaviour when mixed with mortar or concrete ingredients. This question did not receive in the early period of Portland cement so much attention as it does now, and indeed the mechanical facilities required for its accurate performance were not then accessible. It has now become a matter of much importance, and the cost incurred in the realization of an almost impalpable powder is rendered more onerous from the fact of the cement being of a denser character and increased specific gravity. The best values are obtained from the heavy cements, and their fineness of texture has an important bearing on the tensile results. It is therefore necessary to be careful that where heavy cements are obtained, they should be finely ground, otherwise the advantages of their most economical properties are not realized.

It is true that experiments made by Mr. Mann with a finely sifted cement, capable of passing through a mesh of 0·0231 of an inch in diameter, exhibited very fluctuating results, and did not reach in tensile value a coarser cement, 15 per cent. of which was rejected by a sieve of the same gauge. The cement briquettes were broken after seven days' immersion in water, and sustained a tensile strain of the fine cement 378 lbs. per square inch, and the coarse cement 408 lbs. per square inch. The difference is not, however, of much consequence, seeing that the cements were of different brands, and therefore no reliable data can be established through an experiment of this kind. In another series of experiments the same experimenter arrived at interesting results, showing that the coarse cement possesses a much better capacity of adhering to cement than sand. Fine sifted cement and sand mixed in equal parts by weight, gave in twelve tests an average of 217 lbs. per square inch, and fine sifted cement mixed weight for weight with coarse cement gave 328 lbs. per square inch. Such a result is not at all astonishing when we regard the favourable lava character of the coarse cement and its capacity through its porosity of adhering or rather offering favourable conditions for the adhesion of the fine cement. In such a position we should not expect to have other than these results, but the following experiments show conclusively that there is no advantage in an addition of coarse cement for mortar. There were four qualities (meaning thereby their degrees of fineness tested by the sieve previously referred to) of cement, viz.: 1st, fine; 2nd, 10 per cent. of coarseness; 3rd, 15 per cent.; and 4th, 20 per cent., mixed in equal volumes with one quality of sand. Breaking strain applied after seven days' immersion in water gave the following breakings per square inch, on an average of eight samples.

	1.	2.	3.	4.
Lbs. per square inch	207½	215	212	220

Mr. Mann says on this subject: "In many instances he found that mortar, consisting of 1 part of cement to 3 parts of sand, the cement containing 25 per cent. of coarse particles, possessed little more than one-half the tensile strength of mortar gauged in the same proportion with fine sifted cement, the age of the samples being four weeks."

This test of coarseness is one of much importance, and derives its greatest value from being able, when pushed far enough, to detect the quality of irregularly burnt cement. Properly burnt cement should consist of what is technically termed mild clinker, sufficiently vitrified to produce the desired weight or quality. When this is accurately performed, the cement produced will, whether in the condition of fine or coarse powder, be exactly similar in chemical value. It is otherwise, however, when a blended parcel of heavy clinker and "yellow" (light) is passed through the reducing machinery. In every stage of its reduction the fine light powder interferes with the economical reduction of the clinker, and in fact interposes between the coarse and harder particles, rendering their perfect reduction practically impossible. A cement of this character would give the unsatisfactory results obtained by Mr. Mann.

Lieutenant Innes, R.E., in his tables published in Mr. Grant's *book*, throws much light on this part of the question. The table is headed "Results of Experiments on the Effect of the Coarse Particles ($\frac{1}{50}$th of an inch and upwards in diameter) of Heavy Portland Cement."

"The same cement was used for the whole series, being first all sifted and then re-mixed in the proportions shown. As received from the manufacturer it weighed 117 lbs. per bushel, and contained more than 30 per cent., which failed to pass a sieve of 2500 meshes per square inch.

"The percentages are all given by weight.

"The ultimate chemical composition of the fine and coarse grains was almost identical."

Neat Cement in Water.

Fineness.	Tensile Strength in lbs. per square inch.	
	3 months.	6 months.
Passed through 2500-mesh sieve	464	492
Ten per cent. rejected by ditto	402	508
Twenty ,, ,, ditto	499	510
Thirty ,, ,, ditto	507	515
Forty ,, ,, ditto	500	520
Fifty ,, ,, ditto	490	525
Sixty ,, ,, ditto	512	527
Seventy ,, ,, ditto	489	535

The experimenter observes:

" Had it not been expected that the maximum strength would have been reached with a much smaller or no proportion of coarse grains, higher proportions would have been tried, but there would have been considerable difficulty in gauging the coarse cement freed, or nearly so, from the fine, as it was as harsh and incoherent as so much gravel."

To show the injurious influence of coarseness, the same careful experimenter investigated their different values, and found that cement received from the manufacturer weighing 116 lbs. per bushel, after rejecting 20 per cent. by a 1296-mesh sieve weighed only 102 lbs. per bushel, and when still further sifted through a 2500-mesh sieve, rejected 30 per cent. the bushel, then only weighing 99 lbs.

On submitting these three qualities of neat cement to tensile strain after three months' interval, the following results were obtained:

        No. 1—504 lbs. per square inch.
        ,, 2—457 ,,   ,,
        ,, 3—449 ,,   ,,

When mixed with 1 of sand to 1 of cement,

        No. 1—286 lbs. per square inch.
        ,, 2—324 ,,   ,,
        ,, 3—377 ,,   ,,

and with 2 of sand to 1 of cement

> No. 1—117 lbs. per square inch.
> „ 2—212 „   „
> „ 3—250 „   „

3 of sand to 1 of cement

> No. 1—103 lbs. per square inch.
> „ 2—143 „   „
> „ 3—173 „   „

The same cement neat at six months

> No. 1—527 lbs. per square inch.
> „ 2—492 „   „
> „ 3—486 „   „

1 of sand to 1 of cement

> No. 1—350 lbs. per square inch.
> „ 2—424 „   „
> „ 3—439 „   „

2 of sand to 1 of cement

> No. 1—229 lbs. per square inch.
> „ 2—281 „   „
> „ 3—323 „   „

3 of sand to 1 of cement

> No. 1—151 lbs. per square inch.
> „ 2—173 „   „
> „ 3—228 „   „

The cement in these various qualities exhibited at three and six months results against the finest; but when we examine the breakings of the mixtures with sand in various proportions, the fine overtakes the coarser, and at the end of three and six months is very much its superior. This series of experiments goes far to prove that to ascertain the constructive value of cement you must use it in combination with sand. It is the adhesive value we want, and its greatest strength in that direction is found when most finely ground.

The weight also of the cement, as proved by these experiments, has much to do with the fineness. Indeed, no compound test can be complete if the test of fineness is omitted.

The discrepancies in many of the early experiments may be traced to this essential ingredient having been neglected.

Mr. Grant in using clean Thames sand in the proportion of 1 of cement to 3 of sand (the cement weighing 112 lbs. per bushel) obtained the following results:

	At 3 months.	At 6 months.
Lbs. per square inch	60·2	103·3

In whatever direction we examine the grinding question, we shall find that it has a most important bearing on the quality, or at least on the economy of the use of the cement. The manufacturers in their opposition to heavy cement and fine grinding, fairly enough say that it is not only the cost of reduction but the extra expense of burning, which in some cases amounts to a third of the whole cost of manufacture. But we think it has been shown that a bushel of finely ground cement weighs less than that coarsely ground, and in that state will accomplish higher results when mixed with sand. Any arbitrary tests, such as those of the Metropolitan Board of Works, are capable of reasonable alteration if the main element in that test of tensile strength is sustained. The cement makers are very much to blame for the existence of the present system of testing. When under the author's advice 400 lbs. per $2\frac{1}{4}$ square inches was adopted, and when the first specification for the main drainage was in proof, a leading maker waited on Mr. Bazalgette and informed him that the test was too onerous and incapable of fulfilment. An appeal to the writer reassured the engineer, and the test was launched with the result, after nineteen years' experience, of being nearly doubled. Mr. Grant, no doubt, has a strong belief in his mind that the improvement in cement during the above period is due entirely to his exertions. But the cause is very simple and can easily be explained. In 1859 the cement trade was in a depressed state, and much competition arose amongst makers for supplying the contractors for the construction of the works of the main drainage, and at prices very much under those now ruling. It is needless to particularize the rush made and the many tumbles experienced in the attempt to secure a prize which really at the time was not worth having, for little, if any, profit was realized by those who supplied the first contracts. There was the honour, however, and in the scramble for it cement was eventually supplied of much higher tensile value than was demanded.

Such a state of things afforded conclusive proof that the makers of the cement were unable to control the quality of their own products, or they never would have supplied cement so costly to produce at a loss, or at least with small profit, as they received nothing for the increased strength.

There is surely a possibility of English cement makers combining to agree upon a common test for their own protection without impairing the constructive value of the cement. The German makers have so agreed, as we have elsewhere shown, at page 289.

The existing specification of the test as now commonly prescribed, is inconsistent, and the element of bushel weight may be eliminated from it, or altered to meet the improved knowledge of the subject. Tensile strength is the all-important desideratum, and while that and the hydraulic member are secured, the others can be modified so as to render the whole more consistent and reasonable.

There is still much ignorance, if the term may be applied, in regard to the process of cement setting; and in the numerous practical experiments before us, little, if any, allusion is made to this all-important subject. That it is due to a series of chemical reactions, the nature and quality of which depend on the preliminary processes of the manufacture, is beyond question, and as they approach the point of excellence, so is the quality of the results obtained.

Indeed, from beginning to end, so much depends on technical accuracy, that unless their performance is similar, no tests can by possibility serve as reliable guides if not made on even terms. Suppose it were convenient, for it is not by any means impossible, to establish an arbitrary analysis so that the chemical product placed in the hands of the experimenter would represent distinct and accurately defined proportions and free from error, the tests in different hands would produce variable results. When in 1859 the author advised Mr. Grant, and put into his hands the then best known form of mould, he little dreamt of the effect that would arise from its inconsiderate treatment. When he resolved to change it, the author showed him that none of the patterns he had selected would give the highest results, but a wedge-shaped mould of the proper form would considerably exceed any of them. All argument, however, failed to dissuade him from his ambitious design of having a mould of his own. He possibly was influenced in this resolve by the story told of the two Highlanders in their contest about the antiquity of their respective clans, one maintaining that his *remote* ancestor was one of the crew of Noah's ark, and the other, to cap his friend, replied that his clan on that voyage had a *boat of their own*. So Mr. Grant went in for a mould of his own, which has tended to disturb all experimenters in their labours. Originally obtained from France, and for many years and now used by English engineers, the old square-headed mould served the useful purpose for which it was intended, and although not in strictness the form from which the highest breakings could be obtained, served all alike. At best these tests can only be comparative, and fine-drawn distinctions about the lines of the moulds or their best configuration are only attended with neglect of the main question, the quality of the cement. No form of mould or ingenuity of testing can make bad cement good, and this excessive refinement in search of the ideal has a disturbing influence on the question. These expensive experiments settled nothing except that the Council of the Institution of Civil Engineers considered their recorder worthy of reward and honour, with the additional privilege, afterwards accorded, of permission to give them to the public. These experiments, made with public money at the expense of the metropolitan ratepayers, *allowed to be published!!* In the preface to these minutes Mr. Grant says:

"The first of the following papers on cement was read before the Institution of Civil Engineers, London, in December 1865; the second in April 1871. Both have been out of print for some time, the first for several years; and as frequent inquiries are being made for them, it has been considered advisable to reprint them, the Council of the Institution having very kindly given their permission.

"Nothing has occurred since these papers were read to invalidate the facts or make it necessary to modify the statements made by the author, who hopes that they may form a safe starting point for anyone who wishes to pursue the subject farther."

The work thus introduced is not exactly a *reprint*, for in vol. 25 of the Proceedings of the Institution of Civil Engineers, at page 79, occurs the following:

"The author desires to acknowledge the obligation he is under to Mr. H. Reid (the author of this book), whose early education and experience as an engineer led him to appreciate the value of Portland cement for engineering works, and whose suggestions and assistance, in the first stages of this inquiry, were of the most important and valuable character."

We hope to be pardoned for correcting this oversight on his part—we daresay the omission was an accident.

There are two points from which the question of cement testing should be examined, as at each of them separate interests, equally entitled to consideration, exist. First, which is perhaps the one entitled to primary examination, is that of the manufacturers. The second, the consumer in all his many phases and conditions, surrounded in general with divers ideas, fluctuating and perverse interpretations of his own personal duty or what is claimed from him by his employers. The conflicting and apparently irreconcilable position of this class we hope in these pages to have assisted to a better reading of Portland cement and its advantages.

First, then, the manufacturer when placing before consumers an article which he regards as capable of meeting the requirements of a specified test, should feel satisfied that the form and conditions of the examination of the cement should be intelligible and inelastic in character. Also, that its performance is within the capacity of his own establishment, and capable of realization through his own agency or those under his supervision and control. Otherwise how is it possible to follow the continually changing whim of the consumer, whose vagaries and performances we have endeavoured to record in these pages. The alterations of the original form of mould may, if uncontrolled, lead us into a chaos of doubt and perplexity. Is the manufacturer to be expected to have a series of machines and their expensive surroundings in his works, and if he cannot keep pace with the *natural development*, is he to be blamed when failure arises? The thing is almost impossible, for the author, although "thinking of it by day and dreaming of it by night," is far from a full and perfect knowledge of the eccentricities at work in the treatment of Portland cement. We will make a short pause here, and show a very recent outcome of this testing mania, not with the desire of disputing its value, but simply to show that every engineer or architect if not ambitious about the mould, is yet desirous of "paddling his own canoe."

The following results of tests, in accordance with specification for controlling the quality of cement used in important concrete works, show a novel treatment of the question.

All Water Tests.

Weight per bushel.	Age of Briquette.	Age of Briquette.	Tensile breaking per square inch.
lbs.	Hours.	Days.	
120	41	..	112
,,	62	..	150
,,	..	8	404
,,	..	11	312
,,	..	15	412
,,	..	32	370
,,	..	62	303
115	43	..	112
,,	60	..	150
,,	..	8	294
,,	..	28	392
129	43	..	50
,,	63	..	60
,,	..	30	375
127	24	..	71
,,	..	32	248
128	26	..	29
,,	40	..	71
,,	63	..	80
128	..	7	212
,,	..	11	226
,,	..	12	265
,,	..	14	276
,,	..	15	314
,,	..	22	290
,,	..	23	387
,,	..	26	316
125	40	..	92
,,	..	3	143
,,	..	11	271
,,	..	14	258
,,	..	15	251
,,	..	26	273
122	40	..	29
,,	..	4	91
,,	..	7	203
,,	..	11	200
,,	..	21	289
119	45	..	71
,,	63	..	120
,,	..	7	144
,,	..	11	227
,,	..	14	228
124	..	3	144
,,	..	7	215

The cement was obtained from first-class makers, and the tests were rigidly and accurately performed during the months of January, February, and March of this year. Their primary purpose, doubtless, was to give early assurance of the quality of the cement, and it may be said, therefore, under considerable disadvantages.

These results are, to say the least of it, somewhat puzzling, for if we examine them from a specific gravity point of view we find that the heaviest (129 lbs. per bushel) at thirty days old gave only 375 lbs., while that obtained from the lightest (115 lbs. per bushel) at twenty-eight days old was 392 lbs.

With such erratic tendencies in the field the manufacturer might almost despair of keeping pace with the continually changing demands, not only on his patience, but a perpetual strain on the resources of his works.

The test against the maker, under whatever circumstances it may be performed, should be with cement only, and no manufacturer should be bound, at all events on the existing system, by any mixtures of sand or gravel, or the numerous and fanciful ways by which they are compounded. He can only control the products from his works, and the mortar or concrete test is a question entirely with the consumer, and one in which he ought not to be called upon to participate.

In Germany the question of testing is receiving much attention, as we have shown at page 289, and an attempt has been made to arrange some definite mode amongst makers and consumers so that the vexed questions of form of mould, machinery, age of briquette, and mixtures should be placed on an established basis. Some such arrangement is necessary in this country, so as to enable all interested in this most important subject to settle down under defined and reasonable conditions of guidance and control. It is now fashionable to become an experimenter, and to endeavour to force upon others the acceptance of certain rules. The great mistake made in this direction arises from an utter disregard, or perhaps ignorance, of the primary principles of the subject. In the ambition to shine forth as successors of the Barlows, Fairbairns, and other eminent experimenters, they forget that the material on which they have to operate is so changeable and unreliable in character as to prevent the possibility of deducing—from such experiments as theirs, at least—any inflexible rules for guidance. Probably before the chemistry of iron making was thoroughly understood, the products of the furnace and forge varied as much as cement does now, but at the time of the experiments of the above-named great authorities its crystallization and fibre were capable of estimation, and the desired quality of both was easily obtained. The natural materials, such as timber and stone, could be selected, and although fluctuating in quality, maintained, under the most rigid examination, sufficient uniformity of character to permit of the establishment of reliable rules by which they might be safely used. In order, however, to show that even these experiments are variable, and exhibit irregular results, we herewith take, first, from Barlow's tables, the following results :

Crushing or Cohesive Force.

(Size of Cubes, 1½ inch.)

	Crushing Weight. lbs.
Portland stone	10,284
Bramley Fall stone	13,632
Granite (Cornish)	14,302
,, (Peterhead)	18,636
,, (Aberdeen)	24,556

Second, from Wilkinson's experiments on the building stones of Ireland :

TABLE I.

	Weight in lbs. required to crush cubes of 1 inch sides.
Limestones, from 90 tests, average	14,000
Sandstones, 83 tests, average	7,790
Slates, from 36 tests, average	13,154
Granites, from 20 tests, average	6,637
Basalts and metamorphic rocks, from 9 tests, average	14,900

Table II.

	Weight in lbs. required to crush cubes of 1 inch sides.
Limestones, from 120 tests, average	15,340
Sandstones, from 84 tests, average	9,353
Basalts and metamorphic rocks, from 16 tests, average	21,970

Mr. Grant included in his experiments the examination of other materials, but as he departed from the usual method adopted by authoritative experimenters, we cannot form a ready comparison, but in Table 42, p. 43, of the published minutes, we find with samples of various surfaces and 1½ inch thick the following results:

	Per sq. inch. lbs.
Granite, on 4½ inch surface	7,865
,, on 9 ,,	9,488
York stone, on 4½ inch surface	6,376
,, on 9 ,,	10,085
Portland stone, on 4½ inch surface	2,882
,, on 9 ,,	4,589
Bramley Fall ditto, on 4½ inch surface	2,434
,, ,, on 9 ,,	4,589

These results admit of no comparison with those obtained through such agencies as Barlow and Wilkinson, the latter, probably, the most exhaustive and accurately conducted series of experiments ever published, not only as regards the cohesive strength of the materials, but from the information as to their density and mineralogical characteristics; the locality from which each specimen was obtained, its capacity of resistance to rupture, in addition to its cohesive value, being accurately recorded. To show with what care and precision these experiments were conducted, we should state that each of the specimens was subjected to a probationary preparation, by being placed in a room of the temperature of a domestic apartment for several weeks, before being submitted to the test.

When the period arrives for making reliable experiments on Portland cement they will have to be conducted in a

336   SCIENCE AND ART OF PORTLAND CEMENT.

much more careful manner than those experiments with which we have been favoured. It is all very well, and doubtless

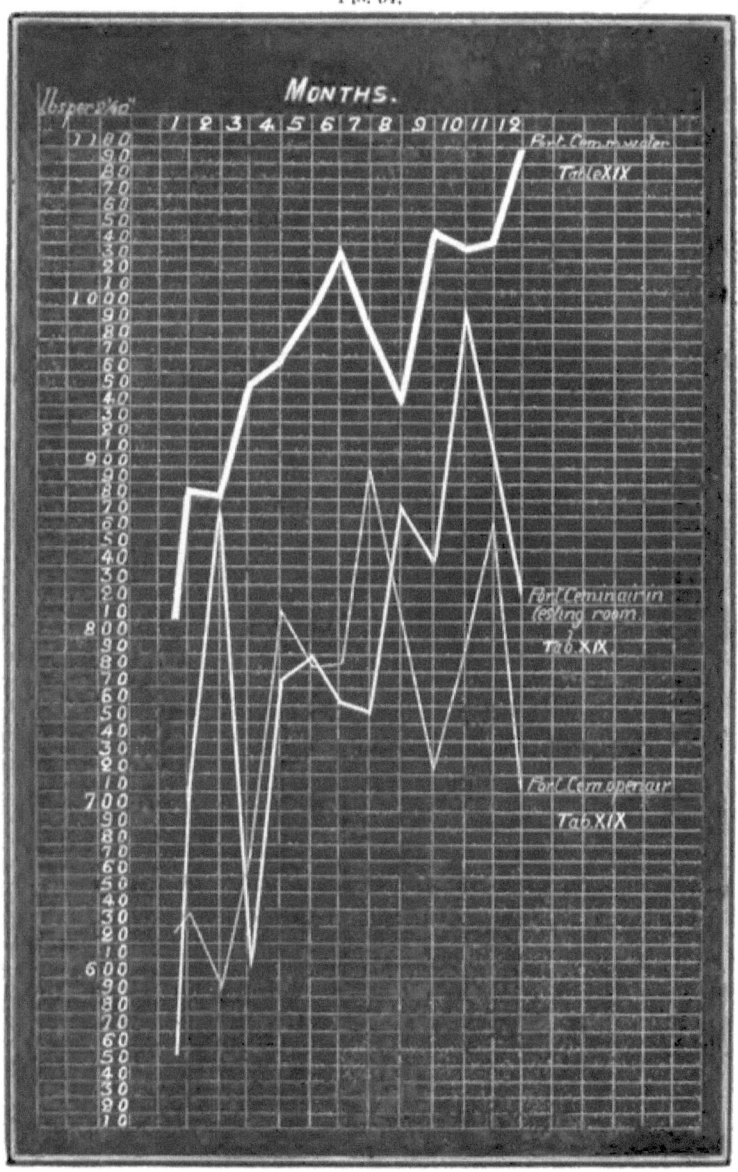

Fig. 54.

instructive, to have tests recorded of such a character as those performed through the agency of Messrs. Grant, Colson, and Mann, but generally speaking they are devoid of the required technical precision, and, especially in those of Mr. Grant, from their promiscuous character. They fail to instruct from the jumbling of facts without any concise or useful deductions being drawn or rules given to guide from their experiences. What can be more bewildering, for instance, than Mr. Grant's Table 19? which, for more facile reference, we have converted into a diagram, as shown in Fig. 54.

The bold line represents the behaviour of neat Portland cement during twelve months while immersed in water; the thinner line indicating the conduct of neat cement in the atmosphere of the testing room; the fine line similarly expressing the value of neat cement in the open air. In the first case is exhibited an improvement during twelve months of 280 lbs. on the $2\frac{1}{4}$ square inches. This maximum was reached in a very erratic manner, and if an estimation had been recorded at the end of eight months only 120 lbs. would have been realized. In the testing-room experiments we find a total increase in twelve months of 260 lbs., and at the end of three months only 50 lbs. increased value, while during its period of maturing much eccentricity of value is apparent. The open-air line, again, in the year gives an improved value of only 80 lbs., while at seven months it reached 270 lbs. From such fluctuating results it is impossible to build up any reliable rules as to value, for this table at least demonstrates either that the workman entrusted to make the test performed his task carelessly and ignorantly, or the cement operated upon was of varied character and quality.

Again in table shown by Fig. 55 we give a diagram of ten years' tests, prepared from a pamphlet published by Dr. Michaelis, of Berlin, who doubtless obtained from Mr. Grant

z

the result of the eighth, ninth, and tenth years' breakings. The eighth year indicates a great falling off in value, which

Fig. 55.

the ninth and tenth years fail to recover. It is to be hoped that no more of these records of questionable value will be published, for their misleading and uncertain character has already created too much scepticism on the possibility of cement improving in value.

Before proceeding farther on this important branch of our subject, we will consider the influence variable quantities of water has on the resulting breakings. This question in the time to come will form a much stronger element in testing than it now occupies. In the use of cement, especially for the production of concrete blocks, the amount of moisture applied to the mixture exerts a marked influence on the resulting compound. At page 368 will be found ample evidence on this point in the case of Buckwell's granitic breccia stone. In the discussions which followed the reading of Mr. Grant's paper at the Institution of Civil Engineers, some valuable remarks were made on the subject of Portland cement and its application, by some of the most eminent members of the profession; and out of the 172 pages of those minutes, 80 pages consist of the verbatim reports of those gentlemen's experience and opinions. The first and most eminent of the English engineers regarded the subject of so much importance as to induce them to give some particulars of their experience in almost every part of the civilized world. The works entrusted to them in the shape of harbours, docks, &c., were more important than those executed under the Metropolitan Board of Works, indicating an appreciation of the merits of Portland cement and its much bolder application contrasting strongly with the limited use made of it by the engineers of that Board. In these discussions the influence of the water used in the concrete and mortar was not overlooked, and we cannot better show its fluctuating and injurious effects than by a reference to Mr. Mann's experiments in that direction.

In the preparation of the briquettes he departed from the usual mode of filling the moulds, and by this alteration enhanced their breaking value. The experiments were limited to seven days' test and nine were made with 5·6 and 8 oz. respectively of water to 32 oz. of cement. The average of these nine tests were for the first 412 lbs. per square inch, for the second 440 lbs. per square inch, and for the third 314 lbs. per square inch. In other proportions of water, viz. 7·9 and 10 oz. respectively to 32 oz. cement, the resulting tests were for the first 367 lbs. per square inch, the second 214 lbs. per square inch, and the third 182 lbs. per square inch; these latter were on an average of six tests. Mr. Mann expresses some doubt as to the continuance of this degradation, and it is to be regretted that he did not prosecute the inquiry a little farther. The weight of the cement is not given. Mr. Colson's experiments assist us here, and show that the deterioration was continuous, and briquettes of six months' age indicated a somewhat lower value than those at three months. In the following table from Mr. Colson's paper is shown the result of his experiments in this direction.

Proportion.	7 Days.	3 Months.	6 Months.
	lbs.	lbs.	lbs.
1 Water to 3 Cement	1·036	1·508	1·544
1 Water to 2½ Cement	804	1·314	1·278
1 Water to 2 Cement	637	1·161	1·094

All these briquettes were kept in water till tested. The remarks describing this table are as follow:—" Shows the tensile strain, at seven days, three months, and six months, of three samples from the same cargo, but mixed with different proportions of water; the average of seven tests being given in each case. The results at six months are not so high as might be expected, the two last in fact showing somewhat less strains than those recorded at three months.

This may perhaps be accounted for from the fact that these tests were made during severe frost."

The most reasonable way to account for the decreased value obtained where an excess of water has been used in mixing up the cement, is from the fact that the soluble silicates have been washed from the mass, and under the best and most careful treatment cannot again resume their normal position, even if they could be recovered. It is this sensitiveness of the most valuable ingredient in Portland cement that leads to so much loss in submarine operations, where the objectionable practice is adopted of placing newly made concrete in position under water. If the concrete after being mixed is allowed to dry, or partially proceed in its initial stage of setting, there is less danger of the cement being acted on prejudicially by the mechanical and solvent action of the water. By such treatment many important works in harbours and docks have been successfully constructed. A quick-setting cement, however, would be unsuitable for such an application or use of concrete.

As in the various references we have made to the results obtained by different experimenters on the character of Portland cement and its various mortar compounds, much diversity of value is exhibited, we purpose offering such remarks as we hope may, if not reconcile them to each other, at least aid in their explanation. To guide us in this object, we will carefully examine the points which primarily indicate the quality of Portland cement, taking the kiln products as our starting point.

First, then, on a careful examination we should find the clinker to be, when of the true quality, a dark greenish mass, thoroughly homogeneous in character and slightly vitrified. When drawn from the kiln hurriedly, a tendency to dust is frequently exhibited, and if but slight, no danger is to be apprehended. If, however, this is excessive, you may be sure that too much clay has been used or imperfectly mixed raw materials permitted. Such a product may be, and frequently is, of high specific gravity, and when converted into powder, in all probability passes the various ingredients of the triple test. The powder is brownish, and feels soft to the touch, reaching its maximum strength at an early stage, failing to improve in value either as a mortar or powder; in fact, the latter becomes on lengthened exposure quite worthless. It was this quality of cement that, before the institution of rigid testing, caused so much trouble and distrust. The cost of fuel required in its manufacture is comparatively low, while its conversion into powder is an easy and inexpensive operation, because in some extreme cases the action of the air has already reduced a large portion of the clinker to powder, or rather dust. This kind of clinker is termed by some workmen "slippy," meaning by the phrase, that it could not hold together. It is due to an imperfect chemical action caused by inaccurate proportions imperfectly ground, resulting in spontaneous disintegration, owing to the physical tension when cooling overcoming the pseudo-chemical combination. Such a quality of clinker is no longer tolerated, and this allusion to it is made for the purpose of removing distrust from the minds of beginners, who too readily condemn materials when such a quality is produced. This result may be brought about by the use of diverse materials under any system of manufacture.

Cement of this character, if put into casks or sacks soon after being burnt, would (more especially if imperfectly ground) swell, and ultimately burst the packages. Great, indeed excessive, fineness can be reached through this quality of cement, and when a very fine powder of this kind is offered, and passes with ease the sieve examination, carefully watch its behaviour in the other elements of the test.

Second.—A clinker of great and almost metallic hardness,

with a black glossy lustre, is the result produced by an excess of carbonate of lime, caused by a wasteful application of fuel. The cost of grinding this quality of clinker is a high one, and the resulting powder under the best treatment is harsh and gritty in character, its colour being of a rather deep bluish grey. These external characteristics are invariably accompanied by a tendency of the powder to fly or blow. The cause usually assigned for this dangerous development is the hydration of the uncombined lime in excess in the raw mixture exerting a mechanical force and displacing the molecules or atoms, thereby preventing their re-arrangement when hydrated. These are usually the consequences of an over-dosed lime mixture, but this sometimes also arises from imperfect comminution of the carbonate of lime. In the former case the cement may, by careful and sometimes lengthened exposure to atmospheric influence, be purged, as it is sometimes termed, and become in a fit condition to be used; but when the latter conditions occur no treatment will render it safe, and the best plan is to reject it utterly. In each of these cases the cement will not pass the water test, and when submitted to this ordeal, its defects are more speedily developed than when exposed to the air only. This quality of cement is usually long in setting when burnt to a heavy clinker, but if light, its behaviour when mixed with water resembles in character that of the lias limes. It develops in this state considerable increase of temperature when setting, and soon afterwards crumbles to pieces.

Third.—The true normal clinker exhibits when drawn from the kiln, a rough lava-like texture, having a colour tinged with a green or bronze glitter when exposed to the clear light. There is a small amount of dust accompanying this quality, more especially when burnt in a high temperature. The practical burner usually examines the clinker of a newly burnt kiln, and if on blowing on a piece of it dust is developed, he regards it as well and properly burnt. If, on the other hand, the clinker is dense and black in colour, as in No. 2, and no dust under the same treatment is forthcoming, he speedily ascertains its character by pounding a small portion of it in a mortar, which after being sifted is submitted to the water test. The powder obtained from the best quality of clinker is light grey, having a slight metallic tinge of colour, and even when finely ground has a moderate roughish feel to the touch. A cement of this character might be safely used from the millstones, although keeping and exposure to the air render it more mellow and homogeneous.

Theoretically we should not have any product from the kiln except clinker, but practically we find that in the best-conducted dome kilns there is almost invariably a considerable percentage of imperfectly burnt material produced. This is due, as in our description of the burning operation we have endeavoured to explain, to the eccentric conduct of the kiln. The accurate quantity of fuel may have been applied, but its imperfect or irregular combustion produced too highly burnt clinker, accompanied with unburnt *pink* and light *yellow*. This is the sort of kiln produce which puzzles and embarrasses the cement maker and tempts him too frequently, sometimes against his own judgment, to blend the mass together, and thus combined place it upon the market. A *mild* clinker is the only cure for this untoward state of things, but that desideratum is not attainable at pleasure under the best-controlled treatment of the ordinary kiln, for there are too many adverse circumstances preventing its perfect consummation.

We have described the leading characteristics of three qualities of clinker, from which we may draw the following conclusions:

First.—The clinker of a brown colour, dusting freely, from

which is produced a brownish and soft powder, indicates an abnormal cement, if not of a dangerous character, at least faulty and weak in indurating capacity.

Second.—The clinker of great density and blackish in colour, unaccompanied by dust, and producing a bluish-grey powder, harsh in texture, dangerous in character, and unfit for use, until after being carefully matured by exposure.

Third.—A true clinker of a greenish tinge, from which a light-grey powder is obtained, capable of being used at once.

Even at this stage we may institute a colour challenge of the cement, which is subsequently further strengthened when the powder is made into sample pats or briquettes. The first quality producing a dingy brown, the second a bluish tint, and the third a fine light grey, resembling when in its highest purity the colour of the best qualities of Portland stone. No exposure to the air will change or alter these characteristics, but on the contrary, when used for mortar or concrete with any kind or quality of aggregates, they exert their distinctive features under all circumstances.

We will follow these three varieties of products to the works in which they may be used, and endeavour to show what results are developed from their several peculiarities.

First.—The cement of a brown colour and over-clayed character exhibits, when used in mortar or concrete, a tendency, at intervals more or less protracted, to disintegrate, not expansively, but apparently from a shrinkage or contraction, indicating that an imperfect crystallization has resulted from a carelessly combined cement. This result is generally misunderstood, and its cause regarded as expansive, notwithstanding the clearly marked distinction between this and over-limed cement. Each has its own peculiar lines of marking, and cannot fail to be distinguished if carefully examined; the one, as in the case of the *brown* cement, almost accurately mapped into innumerable lines intersecting and crossing each other at all angles. This kind of degradation is gradual and extends from the surface downwards until all cohesion ceases. The mass at last becomes pulverulent, and may be blown away. In some cases, where an over-limed cement is highly burnt, there are superficial indications similar in character to those we have described, the difference, however, being that the colour is dark or bluish grey, and the cracks penetrating through a thin skin only, and by exposure of the cement to the air, this tendency to crack ultimately disappears.

Second.—The over-limed cement, resulting in a dark blackish clinker, producing a coarse gritty powder, when used with sand or gravel very soon gives evidence of its dangerous and uncontrollable character. It will not stand the water test, and when introduced into work will after a few days become cracked and *blown*. That is the distinction between the effect produced by the brown cement and this. The cracks are irregular, and proceed from various points, radiating therefrom in all directions. Considerable increase of temperature and volume arises, the effect being produced from an internal source of mechanical disturbance, owing undoubtedly to the existence of free lime becoming hydrated, which in swelling disarranges the imperfect cohesion of the mass. The danger of the disintegration of No. 1 is of a passive character, but the effect caused by the use of No. 2 is of a most active nature and surrounded with the greatest difficulties.

Third, or true Portland cement, under whatever circumstances, maintains its normal form and continues to indurate in a wet or dry position until eventually a perfect crystallization is reached.

Here then are three illustrations of different qualities of Portland cement which may be produced in the same manufactory and from every variety of raw materials.

We shall endeavour to explain what, in our view of the matter, is the cause of the behaviour of the two faulty cements, and if this reading of the difficulty should prove on further investigation correct, it may assist in solving the difficult problem of cement setting.

Perfect crystals slightly expand on crystallization, and imperfect ones shrink; that is to say, bodies in the mass when crystallizing increase in bulk, and if imperfectly crystallized shrink or contract. Indeed, in the case of some metals, as cast iron and certain alloys, such as type metal, much of their value is due to their capacity of expansion when cooled.

The whole process of cement making, from the original mixture of the raw materials to the clinker, is primarily intended to produce a lava-like mass which, when reduced to powder, possesses the property of re-arrangement on the application of the necessary moisture; in fact, assumes the capacity of crystallization, and, although but slowly, ultimately attains great hardness. Those bodies, however, which accomplish their induration in the longest time become eventually the hardest.

From these fixed and unchanging laws we may fairly deduce that Portland cement partaking of the character of a crystallized mass is subject to their government and control. Hence we find in the case of a perfect cement the expansion incidental to its condition exerts a binding influence on the materials with which it may be associated, and according to their quality and fitness will be the result produced either in a strongly compacted mortar or concrete. The mechanical force of the cement exerted during crystallization penetrates into every void space or pore and keeps each particle of sand together in firm embrace in proportion to the nature and extent of its energy. Light cement and brown cement will also, and generally with greater initial energy than the true normal cement, accomplish the process of concretion in a much shorter time, and apparently with greater success. It fails, however, to maintain its hurriedly acquired position, and owing to the imperfect nature of its crystals shrinks and in its degradation drags with it the aggregates with which it was in combination.

It is generally found that cement mortar joints exhibit, when an attempt is made to separate them, a tendency to resist fracture, and when good cement has been used, the materials which it had joined breaks outside the line of mortar. An examination of a joint of this character exhibits the most intimate connection between the binding agent and the brick or stone, so much so, that when these materials are porous, there is some difficulty in detecting the exact line of contact between them. A piece of surface plastering with cement mortar does not display an equal amount of energetic adhesion, but is more readily detached from the wall, and on comparison with the mortar joint is found to be less compactly crystallized and deficient in indurating value. These two results are influenced in some degree by the surroundings of their position, for the joint has the advantage of the superincumbent weight, rendering it thereby more dense and compact, helping it also to resist the mechanical force of the crystallization of the cement, and thereby intensify its compacting influence on the sand with which it is combined. The plaster mortar has no such adventitious assistance, and the value of its junction with the wall is limited to the adhesive action of the cement, while the mechanical influence of the crystallization is dissipated by the absence of the necessary resistance to its natural development.

The silicazation theory of a cement mortar being benefited by a chemical influence from its associated sand or gravel is no longer tenable, neither can we put much if any value on the recarbonating process of the atmosphere. The former question was finally, in our opinion, set at rest by the accurate investigations and analyses of Mr. Spiller, fully described in the author's book on concrete (2nd edition) page 56. The air action has long been a favourite mode of explaining the hardening of old mortar, and, to a very limited extent, no doubt, influences that operation when the conditions for its perfect performance are forthcoming. We must, however, look elsewhere for the results so speedily and perfectly attained in the case of Portland cement, and hope that our observations may ultimately assist in the correct reading of this hitherto unexplained riddle.

CHAPTER XX.

CARELESS USE OF CEMENT.

There is none of the material used in building which requires a more accurate knowledge of its properties than Portland cement. Nevertheless, its peculiarities, and indeed its very chemical and physical characteristics, receive but slight attention. During the early period of its introduction, and when but little heed was given to its quality, no great damage arose, owing to the limited area of its application.

It is, however, different in the present time, when the great impetus given to its use by the application of concrete for building purposes renders an accurate knowledge of its properties more imperative. It is unsafe for the builder to repose too confidingly in the name by which it is sold, or the merits which its purveyor represents it to possess. Let those who use it exercise the simplest precautions, trusting alone to their own judgment, and not, as is too frequently the case, feel satisfied with the imaginary security of the name impressed on the casks or sacks, unless covered by some reasonable guarantee.

The necessary knowledge to ensure the application in a general way for the exercise of the required safeguards is not difficult of attainment, and any ordinarily intelligent workman might readily be instructed in the mode of performing the required tests. Builders have usually a fair discriminating sense of the value of bricks, stones, timber, iron, and other less important materials with which they operate. The small additional trouble to acquire an equally sensible appreciation of cements would be well repaid by the confidence—in dealing with a material so imperfectly understood—which such knowledge would impart. If proper discernment was exercised we should hear less of the discontent and anxiety about cements and their behaviour under various conditions. When the builder is free to exercise his judgment and unfettered by the sometimes ridiculous specifications of the architect or engineer, there can be no excuse for his neglect of duty in satisfying himself of the quality of that for which in his contract he is primarily responsible. The many serious blunders and evasions resulting from the pernicious habit of controlling the selection of materials, and dictating from whom they are to be obtained—often in utter ignorance of and regardless of their quality—is too notorious. In some measure —indeed we may say in a great measure—this practice is gradually increasing the recklessness of builders and contractors, placing them at the mercy of the specified makers, who may impose such prices as satisfy their greed, leaving the consumer to obtain by evasion the profit otherwise unattainable.

Let the contractor who undertakes a duty perform it in an honourable way, untrammelled by unreasonable restrictions of any kind. If an eligible conductor of an undertaking can be entrusted with its execution—and such eligibility should be the only test—give him full scope for exercising his practical experience and let him be charged with the due responsibility of his task.

It is not intended to insist upon the infallibility of ordinary tests for challenging the quality of cement—for with the utmost care failures arise—but only in a modified degree.

In extensive engineering works, where large quantities of cement are consumed, proper safeguards are of necessity provided against the possibility of cement of a dangerous character being used. In the prosecution of such work, the danger of using bad cement may be regarded as but slight. It is in the less important operations of house building that the dangerous use of cement is more to be apprehended, from the impossibility of instituting the required precautions for safety. The comparatively modern fashion of erecting by frames and machines monolithic concrete structures, increases the danger, from the hurried and often careless way in which this kind of work is pursued. Much evil notoriety is thus attached to this kind of work, and consequently Portland cement receives the largest share of blame, as it is convenient, when blunders have been made, to catch readily at the first excuse which offers. Therefore cement falls in for its due, and unfortunately experience clearly shows that there is much cause for blame on the part of some cement manufacturers. We purpose showing in this chapter the evil effects which have arisen from a too confiding reliance on a Portland cement of inferior character. This is one of many cases which have come under the author's observation, and it is selected as a favourable illustration of the dangers to which contractors are exposed when they neglect ordinary precautionary tests.

The accompanying woodcut, Fig. 56, represents the condition of a group of concrete cottages, which tumbled to pieces suddenly and without warning of any kind.

The builder, of considerable experience in this kind of work, purchased the cement direct from the manufacturer, and the labour of building was carefully performed by competent workmen. Fortunately the accident occurred after the men had left work, and so no loss of life took place. There were no circumstances of an exceptional character, such as heavy rainfall or frost, to cause the disaster, and indeed the builder possessed too much knowledge of his business to look for the cause elsewhere than at its true source. An examination of the cement at once convinced him that

CARELESS USE OF CEMENT. 353

Fig. 56.

the fault lay there, and he immediately asked the manufacturer of the cement to compensate him for his loss. The claim was resisted, and eventually it resulted in an action at law for the recovery of the amount claimed. The author, in his capacity of an expert in such matters, was asked to give evidence at the trial, which required him to institute examination and analysis of the sample of cement sent for his inspection.

The preliminary examination showed that the sample differed from a sound Portland cement, in containing a large amount of carbonic acid and impurities; the latter to the extent of 14 per cent., 3 per cent. of which was unconsumed coke. The remainder, on being subjected to microscopic observation, exhibited the appearance of calcined shale or slate. The cement, when submitted to the ordinary mechanical tensile test, broke at about 100 lbs. per square inch.

ANALYSIS OF THE CEMENT, BY W. F. REID.

	Wet.	Dried.
Lime	34·56	48·98
Magnesia	1·44	2·04
Manganese	0·73	1·03
Alumina and oxide of iron	11·26	15·96
Alkalies	1·93	2·73
Silica	20·63	29·24
Water	15·05	
Carbonic acid	11·19	
Coke	3·21	

It will be seen, by a comparison with a good average Portland cement of the following analysis, where the weak and dangerous points of this faulty cement lay. The cement, in addition to its more dangerous ingredients, was imperfectly ground.

ANALYSIS OF A GOOD PORTLAND CEMENT.

Lime	60·05
Magnesia	1·17
Alumina	10·84
Silica	24·31
Alkalies	1·54

There is indicated between the two analyses a difference in lime of 10·20 per cent., and in the clay of 12·27 per cent., the former being deficient and the latter in excess.

The above comparison shows that the sample of spurious cement had a large excess of argillaceous matter, which was practically increased by the deficiency of lime. Had this argillaceous matter been in a state of chemical combination, as in the true Portland cement, the sample examined could never have set at all; its not having been so combined can only be ascribed to two causes; the one imperfect or incomplete calcination; and the other, a subsequent mechanical admixture of burnt shale. We are bound, however, to ascribe the cause to the former, owing to the existence of carbonic acid and unconsumed coke.

In the manufacture of this false product, some argillaceous substance other than clay had been employed; the schistose or laminar structure of that substance, as well as its chemical composition, lead to the supposition that it was either shale or some allied form of argillaceous schist.

It should be stated that the work was executed in the early part of the year, and this doubtless influenced to some extent the resulting catastrophe. No precaution, however, could have secured a satisfactory result with such a quality of cement, for sooner or later its inherent defects would have been developed, and it is a subject of congratulation that the cottages were not finished and occupied.

In the end, the action at law was compromised on the eve of trial by the manufacturer of the cement paying the builder a sum he consented to take in settlement of his claim. An investigation would have exposed the quality of the cement, and have damaged the reputation of the maker. The builder may not be regarded as quite blameless, although he reposed confidence in the quality and the vendor from having used it frequently before for similar and other purposes.

Probably the price which was paid in this case, a comparatively low one, had some influence in its selection. A very superficial examination, guided by the necessary intelligence, would have sufficed to show the faultiness of this cement. Its colour was a dirty or brownish yellow, rough in texture, and exhibiting to the naked eye black and white specks; the former from the unconsumed coke, and the latter from carbonate of lime. These characteristics do not belong to a true Portland cement, which is uniform in texture and of a deep grey colour tinged with green. Failing the exercise of these simple precautions, it would only have been necessary to make up a small sample pat, in a plastic state, when the feeble energy of its setting properties would have been apparent, or its immersion in water at once proved its deficiency in hydraulic capacity.

In all probability the maker was ignorant of the faulty character of the cement, and supplied his customer in good faith, trusting perhaps to its being challenged, if bad, at the point of consumption. It is to be hoped that such blind confidence in each other as here displayed, is not of frequent occurrence, for the neglect of duty on both sides might have been attended with serious consequences.

When accidents or blunders such as that described occur, it is natural to suppose that all the parties concerned will studiously avoid publicity, which would seriously affect their reputations as builders and cement makers. There are, however, cases notwithstanding the utmost attempts at secrecy, which reach the columns of the public press, and it is then somewhat amusing to observe the amount of knowledge "after the event" which becomes developed and then dedicated to the interests of the public safety. These ventilations, as they may be called, exhibit in their discussion marvellous discriminating powers in the detection of the spurious article. One correspondent of the 'Builder' says that he is competent to judge of the quality of Portland cement by the touch or finger manipulation thereof. Such prescience is, in the opinion of the writer, more likely to encourage than prevent the sale and use of improper cement.

In thus lengthily commenting on the case of misfortune so prominently and reliably brought before us, we would earnestly remind all builders and other users of Portland cement, that the best guard against the chance of accident is the exercise of a very small amount of ordinary care on their part. Accidents are not all caused by bad cement, but many are also due to its ignorant and careless treatment—ordinarily to carelessness, but sometimes also from a too free admixture of improper sand or gravel.

The general increase of cement structures has led to a much better knowledge of cement and its properties on the part of the workman, and he no longer blunders in the dark as to the quality of the material which plays so prominent a part in modern buildings. It will be of the highest importance to all concerned in building operations, when the men who guide the manipulation of the materials become truly sensible of the importance of their vocation, and the necessity for technical intelligence in those by whom it should be guided.

The illustration with which we have so opportunely assisted the discussion of this division of our argument should be regarded in the light of a beacon to warn others of the dangers attending a too implicit reliance on any cement, however plausibly presented and extolled. The shape of a cask, or the colour or configuration of a label with which it may be distinguished, should no longer influence or regulate the consumers in the choice of the cement. The risk of using bad cement is altogether his own, and to his judgment alone should he trust for a safe guidance in the matter.

We feel confident that what has been set forth in this chapter will not deter anyone from prosecuting farther cement operations, but on the contrary assist in stimulating a more healthy discrimination in the selection of cements, and the exercise of the easy tests by which their safe employment can be secured. The reasonable and unavoidably straightforward explanation attending the loss and annoyance in the case before us only shows too clearly how little judgment or precaution is requisite to avert like consequences.

There is probably some excuse—at least it may be suggested—for those builders in remotely situated districts for attempting the improper use of cement owing to its high cost.

Until it can be shown that Portland cement of a high character is easy of obtainment outside the charmed circle of London production, many of the existing difficulties are likely to continue. The cost of transit is so high that at the points of consumption the original price is sometimes more than doubled. Hence the necessity for limiting its use, or compensating for its costliness by adulterants of various kinds. A very common practice is to mix the cement with lime mortars, and, indeed, the writer has more than once in his experience met with country builders who regard such a combination as indispensable. It develops, in their judgment, the properties of the cement, and admits of more general application for the various purposes to which it is applied.

In the execution of the Main Drainage Works under the Metropolitan Board, and long after the establishment of the tests for challenging defective or improper cement, an exciting incident occurred at the Crossness pumping station.

The reservoir for the reception of the sewage of the southern suburbs of London during the interval between the tides, when it was to be discharged, was of considerable extent, and an elaborate and somewhat expensive plan was prepared for its construction. The piers from which the groined arching was to spring were numerous, and a large number of them had been built, when it was found that they became distorted and incapable of receiving the weight which they were intended to sustain; in fact, were dangerously disjointed. Mr. Grant, who had charge of the works, and all under him, were puzzled at this peculiar and unlooked-for development of a combination of bricks and cement, more especially in this particular part of the work, for which they expected great credit. Of course the cement was at once blamed for the misbehaviour of the piers, but before condemnation was pronounced, the author was called in, and met Mr. Grant, the contractor, and the cement maker, on the ground. After a large amount of practical (as it is generally called on such occasions) observations by all interested, the author was asked for his opinion, which, indeed, had been formed at a glance. It was not the fault of the cement, for the entire cause was due to the action of the bricks (Suffolk facing), which had become saturated with water, and thus developed a faulty process of manufacture. An excess of chalk marl had been blended with the earth from which they were made, and the free lime, imperfectly mixed, became hydrated by the rain, and produced the distortion of the piers. They had to be pulled down, and it was fortunate that it was discovered before the arches were turned, otherwise disastrous results would have ensued.

A somewhat similar faulty development occurs in bricks made from the lias clays, and generally speaking the blame is laid to the mortar. This is unreasonable, as a careful examination would prove that it was owing to the influence of the iron pyrites in the clay from which the brick was made, and which by the action of heat when burnt, favours an eventual distortion similar to that exerted by unslaked lime under like circumstances.

These observations will, it is hoped, lead to a more sensible examination of the action of limes and cements with bricks and stones, for there can be no doubt that in many cases the fault is ascribed to the cement, when a more intelligent examination would have traced the cause of damage to the right source. While engineers and architects holding leading positions continue so indifferent to the question of cement and its qualities, we cannot wonder at the almost universal apathy displayed by the less prominent members of both professions. Indifference is too mild a term to express the existing condition of the constructive mind, for it really amounts to an utter and obstinate resistance to the inception of all information on the subject of cements and their peculiarities. While such a state of things continues, the writer will not be surprised to hear that an eminent architect has suggested or specified a mixture of Portland cement and common lime mortar for important works, or to hear of an equally eminent engineer insisting that all cement should pass through a mortar mill in the same manner as that generally adopted in the preparation of common lime mortars. Either of these two novel applications of what ought to be well-understood materials, would indicate with too much accuracy the state of cement intelligence in high places. There is unfortunately a tendency at the present time to diverge from the path of safety—one now tolerably well beaten and defined with the best records of experience in the search for novel combination of mortars. When such mixtures are reached, or rather sought after under the guidance of chemistry, little danger is to be apprehended, but when undertaken as " leaps in the dark" after something fresh and novel, they are surrounded with danger.

In thus advancing the claims of chemistry on the constructive professions, we run the risk of being challenged, in opposition to our argument, that it failed to avoid the disaster of the Houses of Parliament. In that case, however, the architect was over-anxious to secure workable stone, and the chemistry placed at his service was not guided by practical reasons or experience, entirely overlooking the character of the atmosphere in which the stone was to be used. The purity of the atmosphere in the locality of the quarry where the durability of the stone was tested, differed greatly from that of Westminster. The Egyptian Pyramids even would not long resist the degrading influences of such an atmosphere, for their endurance and stability is entirely owing to the dryness and purity of the air in the locality of their erection.

Wren and Smeaton were satisfied to accept such materials as were well known, and St. Paul's Cathedral and the Eddystone Lighthouse are the inheritance to posterity of what great and earnest men can accomplish.

Railway bridges and tunnels, docks, harbours, and mills, warehouses and dwellings tumble about and create but little sensation in these days. But if St. Paul's Cathedral in any of its noble and magnificent parts failed, or the Eddystone Lighthouse succumbed to the elemental strife with which it is surrounded, civilization would sympathize with what would be regarded as England's national disasters.

CHAPTER XXI.

THE VARIOUS USES TO WHICH PORTLAND CEMENT IS APPLIED.

GEOLOGY, during its long and patient investigations, aided by the allied sciences of zoology, botany, and palæontology, brings familiarly home to our minds the history of the earth, its many strange changes and developments. In the establishment of its true and now well acknowledged principles, the varied mining industries have been benefited when by its teachings they were honestly guided, and disaster and ruin created when the lessons which it propounded were disregarded and set at naught. Mineralogy, the natural sequence to the knowledge of the formation of the rocks, continued the investigation, and established in extended lines their physical character and other striking peculiarities. But it was not sufficient to distinguish the various rocks by their colour, fracture, crystallization, and other established distinctions, for the most intelligent application of the rules laid down at best only succeeded in the obtainment of a superficial knowledge of the mineral under review. Chemistry in its all-searching power naturally followed, and there was then no longer any doubt regarding the value of the minerals and their usefulness to the arts and manufactures. Geology described the architecture of the rocks, mineralogy classified their varieties, and chemistry finally tested their commercial value.

It is from this triple association that the art of Portland cement derives so much aid; for through their combined agency we are guided in the selection of the rock, can estimate its superficial character, and ultimately prove its inherent properties.

In artificial stone preparations, not only are we in a position to rectify and avoid the shortcomings of nature, but we are also enabled to point out and determine their cause and effect. The cementing agent in the rock can be proved to be either calcareous, argillaceous, or siliciferous, and the various grains or pebbles which they hold together can be in like manner estimated. Proceeding farther, we can value the atmosphere of their intended site when placed in buildings, and ascertain its suitability for their permanent erection and duration. With such knowledge and its intelligent application, the concrete and artificial stone-maker may not only imitate the best and most durable rocks, but surpass them in all the modern requirements for which many of them are unsuited. We have shown that other cements, such as that made from carbonate of baryta, excel, for some purposes, the best Portland cement, and attain a point of induration to which the most improved and carefully made production cannot hope to reach. Soluble silicates, which are now comparatively expensive, may be by improved systems of manufacture placed within our economical command, and thus again another most valuable agent brought to bear on the production of durable combinations for constructive purposes.

The quality of building stones can readily be tested by an examination of their colour and texture. It will be found that when they are subjected to accurate observation, uniformity of colour and fineness or evenness of their grains or particles indicate the greatest durability. Indeed, so reliable are the deductions which may be drawn from an intelligent application of such a test, that it would not be difficult to construct a diagram which would measure at a glance the separate value of each stone, ranging from the durable Portland to the irregularly and loose grained gritstone.

In discussing this department of our subject, we find a difficulty in particularizing the various purposes for which Portland cement in some form or other is used. The growth of its use has been influenced by the improvements in its manufacture. Originally intended as a plaster or mortar in connection with house-building purposes, its success in that direction gave but faint indication of its ultimate usefulness. When engineers became impressed, not only with its useful properties, but the reliability of its quality, its sphere became enlarged to such an extent and with such rapidity, that for some time the demand exceeded the supply, so much so in some cases as to interfere with the profitable progress of important works at home and abroad.

The harbour and dock works of this and other countries may be regarded as the more important and extensive application of Portland cement. Indeed, many of these valuable structures were, before the unhesitating acceptance of this useful material, impossible. Drainage works in cement concrete are not yet of general occurrence in London, which may be in a great measure due to the limited use of concrete by the engineers of the Metropolitan Board of Works. In these expensive works, large sums may be said to have been wasted in needless details, and in a quality of work almost too costly even for metropolitan external building. The construction of the several Thames embankments offered excellent opportunities for a wide and useful, as well as economical display of engineering skill, but the temptation of using the expensive granite prevailed. The lesson which can be so easily read at Waterloo Bridge, failed to attract enough thought and consideration to the fact of the perishable character of granite, placed in a London atmosphere. Our Houses of Parliament, again, should be a warning to all engineers and architects to study by the light of chemistry the nature of the materials they intend to use, and the climate of their intended erection.

The numerous examples of concrete work in the United Kingdom afford ample opportunity of studying the subject in its most comprehensive as well as minutest details. Although the river Liffey works are not the earliest important application of concrete blocks, they certainly to the present continue the boldest and most daring in their conception and execution. The embankment and walling of a tidal river, busy, if not crowded with steam and sailing vessels, in such a manner as that executed under the direction of Mr. Stoney is highly creditable to the scientific and practical ability of that engineer. In the direction of river improvement, concrete blocks after this remarkable example are capable of the most facile execution. The harbour works at Douglas, in the Isle of Man, afford a good example of concrete work, built under the direction of Sir John Coode, who has perhaps the largest experience in such works amongst English engineers. Built in a highly exposed situation, these works withstand the severest gales, and, with but slight exception, without injury or displacement of the blocks. As an evidence of what example induces, it may be stated that previous to the commencement of these and similar works in the Isle of Man, Portland cement or its properties were scarcely known. At present a large consumption of the article has arisen in the execution of other works of a less important but useful kind.

The extensive Government works at Dover harbour have been for many years constructed of concrete blocks. It is now a question whether engineers should not direct their attention to the necessity of increasing the size of the blocks to be used in more than ordinarily exposed situations.

During the violent storms of the past winter, several weak points in the existing system of building dock and harbour walls have been developed. There is practically little difficulty in at least doubling the present size of such blocks, and in some situations they might be made equal to those used in the works of the river Liffey. When favourable materials exist—and in most harbour works the obtainment of suitable aggregates is a matter of little difficulty—the size of the blocks need only be limited by the mechanical means accessible for their economical handling and deposition. The building up of such blocks is not necessarily done by expensive or skilled labour, at least where intelligent supervision exists, and no work of this kind should be undertaken in its absence. Not only should the director of harbour or dock works have a thoroughly practical knowledge of construction, but it is now required that in addition he should possess such technical acquirements as will enable him to estimate the mineralogical value of the stones he is to use, as well as a capacity to challenge a faulty cement. Guarded by these necessary qualifications, he would be the more confidently prepared to encounter the difficulties of his position, and be less fearful (because of his knowledge) to diverge from the beaten path of routine and dogmatism. How many of the failures in constructive operations are due to ignorance and prejudice. Many an old and valued clerk of works of bygone days had seldom more than one idea. If originally a carpenter or joiner, everything he did or thought had a strong dash of *timber* about it; or if originally a mason or bricklayer, believed only in bricks and stones. Limes and cements, except their various superficial characteristics, were Greek to him, and if they set quick and made a fuss in doing so, he was satisfied. All these ancient peculiarities have now disappeared, and there are clerks of the works who really do believe that slow-setting cement is best, and when it is placed in their hands to use, thoroughly appreciate its most valuable properties.

Novelties, or a departure from the beaten path of what is called practical experience, generally meets with silent and dogged opposition of a somewhat negative and undemonstrative character. This may perhaps be regarded by some as a safety valve or gauge test through which all innovations and improvements must or should pass before obtaining the necessary healthy recognition which will save them from sinking into oblivion. Numerous instances might be cited of the most useful and valuable inventions receiving from those whom the public regard as the proper censors such "faint praise" as to lead to their being put aside and ultimately forgotten. Portland cement, with its many now generally recognized advantages, has suffered from this form of prejudice, and might have culminated in the same fate if its vitality had been less decided. We now presume that it has passed through the various trials and difficulties incidental to its youth and middle age, and has established its claim to the indulgence usually accorded to those discoveries which have run the gauntlet of fierce opposition, and claimed the prize as a reward for the dangers they passed.

Although an account of the numerous attempts to press the claims of concrete for other than engineering purposes would in our opinion be highly interesting and instructive, we must confine our observations to the first successful application of a concrete prepared on unusually accurate lines, and its more recent use in paving, &c.

It is now nearly thirty years since Mr. Buckwell patented his "granitic breccia stone," and for some years after that date he executed works in tanks, basins, pavements, &c., in various districts of the metropolis. At this period much difficulty was experienced in obtaining the required quality of Portland cement for the purposes of this manufacture, and some of the difficulties in the perfect establishment of this useful concrete combination might be traced to that cause. The primary object of the manufacturer was the command of a true Portland cement not less in weight than 124 lbs. per imperial bushel to bind together in the closest contact such materials as were suited for the purpose. Limestones from the oolitic formation were preferred, but gravel and other similar materials were also used. The process adopted was to incorporate the cement and the aggregates with the least possible amount of moisture by the impingement of rammers acting on the mixed materials in the various moulds. These moulds were of cast iron, having planed faces so as to ensure accuracy of form in the resulting products. The act of uniting the materials was performed in the most careful manner, and the necessary water was imparted by the finely perforated rose of a watering can. So trifling was the apparent quantity of water thus introduced, that the mass had every external indication of rottenness, and bore no superficial signs of its future excellence. There was occasional use made of a soluble silicate when work of more than ordinary excellence was desired, but the cost at that time of the chemical agents prevented its general use.

This "granitic breccia stone" was used in 1851 in making water tanks in Hyde Park and St. James' Park, in the basins of the Crystal Palace, and in the catacombs of the City of London Cemetery at Ilford, in pavements and landings in many parts of the country, and it was also used by the Thames Conservancy for mooring blocks, the engineer of that Board regarding it as superior to roach Portland stone for such purposes.

A pavement 400 feet long and 20 feet wide, laid down in 1853 at Lewisham, in Kent, still exists as an evidence of its suitability for such a purpose; sewage pipes, pillars, and a variety of useful applications in all constructive directions were made, and its ultimate abandonment was due in a great measure to the want of sufficient public acknowledgment of its merits rendering the further prosecution of the manufacture unprofitable, also to the costliness of hand labour. There was, however, sufficient advance made to establish the possibility of equalling by the agency of this manufacture the best known building stone, and also proving that for sewerage purposes it was very suitable.

Mr. Buckwell received numerous flattering testimonials from the engineers and architects under whose direction it had been used for many purposes, after a test of years had sufficiently proved its valuable properties. He was undoubtedly the earliest experimenter who applied intelligent scientific rules for the manufacture of concrete, and doubtless experienced much anxiety and disappointment before reaching the goal of his ambition. The primary object of his efforts was to obtain in the most compact form a concrete with the least possible amount of moisture; in fact, only enough to impart to the dryly mixed mass of stone and Portland cement the necessary water of crystallization. That was at least theoretically what he set out to do, and practically he all but attained his object, for by rapidity of impact in the use of the rammers, he was enabled to so combine all the ingredients (including the moisture) that the stone at three weeks old weighed little less than its original weight when first mixed. Thus indicating an accurate combination of the water, which could only have been absorbed in the process of crystallization, leaving no superfluity thereof, that would have been an element of danger or waste; in fact, everything was conserved, and thus the best conditions of a cemented mass secured. In the earlier trials, much inconvenience was experienced in the moulding process, owing to the difficulty of removing the concreted mass from the moulds. In iron moulds the concrete stuck to them so as to destroy all accuracy in the forms produced. Slate moulds were next used, but owing to the great pressure exerted by the impinging action of the rammers, they were found too weak, and for that reason had to be abandoned, although their texture was favourable for the production of accurate output. Wood moulds were also used, but they were found too weak. Subsequently glycerine was rubbed on the iron moulds and proved advantageous. But the most successful application was soft soap, which overcame the difficulty, and thus enabled Mr. Buckwell to have his moulds constructed of wrought and cast iron of the required strength.

The selection of the materials for this accurately prepared concrete was carefully attended to, and any raw clayey particles or very fine dust were rejected. Slates or hard shales were also looked upon with suspicion, and the more micaceous granites entirely avoided. In some careful experiments made by the Engineer of the London Docks in March 1858, the following results were obtained:

EXPERIMENTS WITH HYDRAULIC PRESSURE OF MATERIALS, BY W. ANDREWS, ESQ., C.E., LONDON DOCKS, MARCH 1858.

	Granitic Breccia Stone Landings.			Calverly Wood Stone.	Ottley Bramley Fall.	Tricketts Bramley Fall.	Yorkshire Stone.	Cheesewring Granite.
	Under 12 months old.		Under 6 mos. old.					
	6" × 6" 3¼ in. thick.	6" × 6" 3¼ in. thick.	6" × 6" 3½ in. thick.	6 inch cube.	6 inch cube.	6 inch cube.	6 inch cube.	6 inch cube.
Pressure in tons to crack	73	73	56	73	62	—	—	—
Pressure in tons withstood before breaking	104	88	81	81	73	52	66	45
Balance pressure in tons withstood from cracking to breaking or going	31	15	25	8	11	—	—	—

In 1845, or probably during the previous year, the discovery made by Mr. F. Ransome of his well-known artificial stone, and about the same time Mr. Buckwell's invention of the "granitic breccia" stone, attracted much attention amongst the more eminent scientific men of that day. Dean Buckland especially considered these two inventions as of the greatest importance, and so impressed was he of their desirability for constructive purposes, that he decided on giving a lecture on the subject at the Royal Institution of Great Britain. His sudden illness prevented his doing so, and the duty was kindly undertaken by his friend the no less eminent and illustrious Faraday.

The lecture was delivered at the Royal Institution on the 26th May, 1848, the notes for which were furnished to the 'Athenæum' by the Professor, as at that time no short-hand reports were taken of these lectures. The extract from that paper is as follows:

"As the artificial stone of Mr. Ransome is chiefly applicable for ornamental purposes, so Mr. Buckwell's invention, termed by him artificial granite, appears exclusively designed to supply the place of blocks brought from the quarry for large works, whether walls of houses or of aqueducts, sewers, &c. Mr. Buckwell uses the following simple process. Fragments of suitable stones (Portland stone, for example) are gauged and sorted into sizes. These are cleaned and carefully mixed on a board with cement in the proportions of 5 parts of large fragments, 2 of smaller ones, 1 of cement, and a portion of water, but the water is in no *greater quantity* than will bring it to the dampness of fresh deal sawdust. This being done, the materials are put into a strong mould to the depth of about $1\frac{1}{2}$ inch at a time; they are then *driven together by percussion*, more materials are now put in; these in turn hammered together till the water has escaped by holes pierced for that purpose in the moulds, and this process is continued until the block or pipe has attained the required magnitude. It is then taken out of the mould, and now found to be so hard as to ring when struck, and in ten days is fit for service. It is affirmed to harden under the influence of moisture, to bear, when moulded in the form of girders, a greater transverse pressure than any rock except slate, and to be only one-sixth of the cost of brickwork.* It will be noticed that the process is characterized by the use of fragments, by the quantity of cement employed (not one-fourth of the proportion used in common *grouting*), and by water instead of fire being made the means of bringing the fragments into close unison. Mr. Faraday then noticed two scientific principles on the success of which Mr. Buckwell's process depends.

" 1st. *The use of water in effecting the approximation of the particles, and the exclusion of air.*—It has been ascertained by Dr. Wollaston (Bakerian Lectures, 1828) that, in order to bring the particles of platina into close contact, it is best to bring them together in water. Where a freshly made road is watered to make the materials bind together, the same principle assists in the result. Having filled a measured glass with sand, Mr. Faraday showed that when the glass was first filled with water and then the sand added with agitation, it occupied less space than when dry.

"2nd. *The effect of percussion in bringing particles together.*— Mr. Faraday noticed that simple pressure will not displace interstitial air or water, but that a blow will. Water contained in a small cylinder of wire gauze was shown remaining in the open network when subject to the pressure of a column of the same fluid, though it freely ran through the meshes when the cylinder was gently struck; on the same principle moistened sand on the sea-shore gives way and

---

* We presume there is some error in this estimate of the cost.

leaves a footmark under the limb that strikes it. In conclusion Mr. Faraday noticed the remarkable fact that the sedimentary matter in sewers, &c., does not accumulate on Mr. Buckwell's granite as it does on glazed pipes."

The most cursory examination of a piece of ordinary cement, mortar, or concrete, will show that a large percentage of the mass is uncombined or imperfectly united owing to the intervention of the air and water spaces; were these voids filled, not only would the density of the preparation be increased, but its strength would also be much improved. The naked eye may readily measure the extent of this injurious interstitial space, but on the application of an ordinary magnifying glass or microscope, the porosity with its imperfection is at once realized.

Although we have thus given prominence to Mr. Buckwell's granitic breccia, it is not from any feeling of disregard to the merits of Mr. Ransome's beautiful invention, but because its claims on our attention are less important to the subject of our argument than the other. We shall, however, give a few particulars of this interesting process, which, like the "Rowley rag" basalt manufacture, described at pages 102, 103, has succumbed to the want of the necessary patronage, although not a commercial failure.

The promise of success with which this manufacture was at first received was not unworthy of its merits, for it promised to meet a want much felt in the architectural profession. It was capable of being moulded into the most elaborate ornaments, and equally suitable for the production of tombstones, paving stones, and grinding and scythe stones. We shall shortly describe the process, for its chemical excellence is well worthy of our notice.

The whole system of manufacture is based upon chemical reaction, and the first stage consists of the conversion of flints into a kind of water glass. This viscid product is afterwards quickly mixed with the determined quantity and quality of sharp sand, which is thus converted into a plastic mass, capable of being shaped to any form. The moulded mass in all its variety of shapes is placed in a solution of chloride of calcium, which is made by pressure to penetrate into every pore. The contact of this solution with the silicate results in an almost immediate decomposition, whereby the silica seizes, as it were, on the calcium, and forms a hard silicate of lime, leaving the chlorine and soda to form chloride of sodium to a limited extent. The silicate of soda which covered or enshrouded the particles of sand before their introduction into the chloride of calcium bath, became saturated with silicate of lime, a most indestructible material. The remaining soluble chloride of sodium is then washed by pressure from the now flinty stone through the agency of pure water, its thorough expulsion being secured by chemical test. After being dried, the stone is suitable for such purposes as may be required. More recently Mr. Ransome has succeeded in dispensing with the chloride of calcium and the various washings which its use necessitated. The prepared natural silica in definite proportions is combined with soluble silica of potash or soda and lime, sand, alumina, or other approved and suitable materials, which, after undergoing an accurate process of the necessary moulding, are allowed to harden, when the action of the silicate of lime ultimately cements it into a solid stony substance. The actions chemically producing the above results seem to be as follows. On the materials being mixed together, the silicate of soda becomes decomposed; the silicic acid, being liberated, combines with the lime, and forms a compound silicate of lime and alumina; while a portion of the soda, in a caustic condition, is set free. This caustic soda combines with the natural soluble silica, thus forming a fresh supply of silicate of soda, which in turn becomes decomposed by a fresh supply of lime, &c., &c. Through this process the whole of the caustic soda is ultimately fixed, doing away with the washing necessary in the original or first process.

Hitherto concrete in mass has not been produced on lines so accurate as those defined by Mr. Buckwell, and indorsed, or at least recognized, by so great a chemist as Faraday. The essential conditions are simple, and, as explained at the Royal Institution lecture, consist of, 1. The discreet and accurate use of water; and 2. The necessity of impact to bring the particles under treatment into the required accurate contact. It will be observed that both the Ransome stone and Buckwell concrete processes aim at and successfully result in an entire absence or expulsion of *all interstitial air or water space*, and produce a firmly compacted mass, which is only possible by their perfect absence. These conditions, the accuracy and advantages of which are undeniable, are not easy of fulfilment under the ordinary circumstances in which concrete is usually employed in such works as floors, roadways, and walling by the monolithic system. In all other structural operations where the preparation of blocks precedes construction, it is quite possible to produce a mass of such a quality as that of "granitic breccia." In the preparation of large blocks for marine works, the aid of machinery of percussion becomes necessary; and not only would the resulting compound be more valuable in quality, but the time at which it could be used considerably accelerated. The porosity of concrete blocks as now prepared renders them liable—owing to their spongy character—to absorb a large percentage of water, making them thus more subject to ultimate degradation by the constant presence of sea water, with which they are unavoidably incorporated. The mass of granitic breccia carefully built not only avoids all superfluity of moisture, but has actually combined with the water needful for its crystallization, and begun or finished that chemical process, according to the time it is allowed to remain before being used, and previous to its being placed in position.

While thus alluding to blocks for docks, quays, and other analogous structures, we will shortly allude to the question of the size these blocks should be. Recent storms, as elsewhere referred to, amply prove that the blocks now used are too small, and in water unable to resist the action of violent and unusual gales. French engineers have given since the time of Belidor much attention to this subject, and have gradually increased the size of their blocks for sea works, basing their calculation principally on the mechanical force of the storm wave of the Bay of Biscay. Some such observations must ultimately guide us in determining on the dimensions and weight of blocks intended to be laid on our coasts, subject more especially to the storm fury of the Atlantic. Geologists have been keen observers of this sea action, and the best illustrations known to us occur in Mr. Kinahan's work on 'Valleys, and their relation to Fissures, Fractures, and Faults,' where, at page 49, referring to the "block beaches" near Arran Island, on the west coast of Ireland, he says:

"The stones forming the 'block beach' are cast up during the winter gales, and some of them are of a considerable size. A little south of Doughatna the following observation was made: 'Great quarrying seems to be going on here during the gales. Blocks $30 \times 15 \times 4$ feet tossed and tumbled about.' And again, half-way between Doughatna and Glassen Rock, there is this note: 'A block $15 \times 12 \times 4$ feet seems to have been moved 20 yards, and left on a step 10 feet higher than its original site.'"

The rock was limestone, and if we take its weight per cubic foot at 170 lbs. for the first block ($30 \times 15 \times 4$), we obtain 136 tons, while the second (15 × 12 × 4) would, by a similar estimate, amount to 55 tons.

Such a formidable natural force is not peculiar to the west coast of Ireland, and its existence in such intensity is likely to be found in many parts of the world, so that engineers entrusted with works in exposed situations should regard the question of sea action as of primary and indispensable importance.

While considering this element of danger from a mechanical force, there is also an ever present chemical agent, which, if not so apparent, is equally dangerous. In different seas, the water is impregnated with variable quantities of magnesia—that of the Mediterranean containing as much as 4 lbs. to the cubic yard of water. Engineers have not hitherto regarded in so serious a light as its consideration demands, either the elemental strife, which may be regarded as mechanical, or the impurity of the ocean water, which is chemical in character, and perhaps more insidious and dangerous than the other. The storm damage is easily estimated, but the insidious effect of chemical wasting or degradation from its unseen and unfelt influence is likely to evade our observation. This sea-water difficulty was one that puzzled engineers of old times, and the damaged and dilapidated piers and harbours of the Mediterranean are the testimony of its power and influence on untrue mortars made from imperfect and ignorantly prepared natural or artificial cements.

Increasing chemical knowledge led to the manufacture of "Victoria stone," under the patent of the late Mr. Highton, who ingeniously took advantage of the beds of natural silica existing abundantly under the chalk deposits in Surrey. He made his concrete preferably from a mixture of Portland cement with refuse or chippings of the Mount Sorrell (Leicestershire) granite, though the same process was applicable to any combination with materials of suitable character and quality. There was nothing essentially different in this process from that generally practised in making ordinary concrete blocks or slabs, and, indeed, no especial precautions appeared to have been taken similar in character to those adopted by Mr. Buckwell. In the one case, however, as in the manufacture of "granitic breccia," the greatest care was bestowed in the accurate mechanical combination of the broken stones and the cementing agent (Portland cement), with such satisfactory results as those which we have endeavoured to describe. On the other hand, as by the method of Mr. Highton, the chemical element was primarily relied upon for the ultimate value of the manufacture. The process itself and its original conception indicated a considerable amount of scientific ability.

Ordinarily architects and engineers have now become satisfied with a fairly compacted concrete mass obtained through the agency of faultless Portland cement in combination with properly selected aggregates. In the process of "Victoria stone" manufacture these conditions were equally sought for and realized, but a step farther was made and the mass thus carefully produced was submitted to a final process, by which was anticipated years, if not ages, of natural induration or crystallization.

The chief aim of the inventor of this process was to obtain, without the aid of expensive heat, a stone surpassing in value that produced under the patent of Mr. F. Ransome. This he successfully accomplished in the following manner. The magic aid of heat was dispensed with, and the fluctuating and uncertain results obtainable through its agency were entirely and successfully avoided. The only fuel contribution to the success of the process was that employed in the manufacture of the Portland cement, although even that may be regarded as having been performed outside of the "Victoria stone" process.

Concrete made under the most favourable conditions exhibits a considerable degree of porosity, and the aim of the inventor of the "Victoria stone" process was to thoroughly impregnate the mass so as to fill up entirely the whole interstitial space. The accomplishment of the task he had set himself was obtained by the introduction of the natural and inexpensive silica obtained from the sources already described. This silica, prepared in a cold bath in which the concrete articles were placed, acted on the cementing agent and ultimately petrified the whole mass. This process is rendered practicable from the cheapness of the natural silica which is combined with soda in the bath, forming a silicate of soda, the cost of which from the ordinary and usual manufacturing chemical source is somewhat heavy. The silica which is absorbed in the silicating process by the cement is easily replenished, and a perfect silicate bath thus maintained at a comparatively nominal cost. Paving slabs, steps, landings, sills, &c., are thus manufactured at a cost much under Yorkshire stone, while the absorptive properties of the artificial material are much less than the natural one.

An accurately performed operation by this process might accomplish such results as would practically amount to a perfect stone capable of resisting the greatest wear and most injurious atmospheric action. The advantage over the natural stones is obviously great, and perhaps the most noteworthy is the accomplishment of accuracy of form, without the aid of skilled labour. Hitherto, and indeed during nearly the whole period of recorded time, great loss arose from the necessity of choosing a stone suitable for easy conversion by the mason or sculptor, regardless of the ultimate effect of climatal action on its durability. The stones best suited to fulfil the conditions of stability and permanency were not easy to work, and indeed, the conversion of some of the best varieties of building stones were under ordinary circumstances impossible. The irregularity of the cementing agent in the natural deposits, and the variable character of the grains or particles of which they are composed, render their use under the most favourable circumstances precarious and uncertain. The damage incurred is sometimes so remote, that those primarily responsible for the mistake practically disregard its attendant penalties, knowing full well that they will have passed away before the day of reckoning arrives. An artificial stone of this homogeneous character, equal to the most severe tests of imperfect laying, and having the capacity of resisting the most vitiated atmosphere, must ultimately attain the position its merits deserve.

There are but few natural rocks capable of withstanding the unavoidable dangers incidental to and inseparable from the various operations of quarrying, chiselling, and bedding or laying. Modern building in its high-pressure condition cannot pause to study the natural bedding of the stone, or take the trouble to see that it is matured or ripe enough for use. There can be no question, however, of the desirability of exercising great care not only in the selection of the stone at the quarry, but the ultimate bedding in its destined site. A very cursory examination of any of the sandstones will show that these deposits have been formed gradually and in very thin layers during pauses or intervals of time, probably not long, but sufficiently so to create a considerable pause in the operation and to leave each succeeding deposit exposed to such injurious atmospheric influence as at that time existed.

A section of some of the sandstones usually employed for building or paving exhibits various laminations, and the division of each, almost invariably marked by a darker line than the stone itself. This is sometimes composed of carbonaceous matter, and tends to the destruction of the stone when exposed to the atmosphere. Such stone, when laid regardless of its planes of lamination and at various angles from the original horizontal or other line of deposition, exhibits in a very short time the evils attending its being so placed. In the pavement, in passing along, we see the *flags* peeling off in flaky pieces and the section exposed, clearly indicating the cause of the fractures, and consequent degradation. Paving flags being irregular in size, and differing greatly in their quality and capacity of resistance to wear, aggravate the difficulty.

The finer-grained qualities exhibit durable properties, but even the best descriptions soon show indications of broken edges, or arrises extending to a gradual widening of the joints, and becoming at last mere water carriers. Granite, again, is unsuitable for pavements, not only from its dangerous slipperiness when partially worn, but also in consequence of the irregularity of its wear. The granite paving of the footways of Waterloo Bridge is a good example of this kind of action. This pavement appears to us in its present condition most dangerous to the pedestrians by whom it is used.

In contrast with the ordinary stone paving from natural sources (excepting asphalte), we would call attention to the condition and quality of several examples of artificial (patent Victoria stone) slabs in various parts of London and elsewhere.

At Piccadilly, near Devonshire House, there has been laid down for some years concrete slabs in size about 6 feet by 3 feet, and their present condition is most satisfactory. Not only do they wear well and evenly under by no means a light traffic, but their appearance is most pleasing, the joints being still perfect; and in short there are no indications of degradation at all. Again, in High Street, Bloomsbury, between the parish church and Tottenham Court Road, the pavement on the east side is composed of concrete slabs about 3 feet square, wearing well, and pleasing to look at. We will refer to another London example: on the southern approach of Blackfriars Bridge, nearly opposite the station of the London, Chatham, and Dover Railway, a small portion of the footway was laid, seven years since, with similar success to the above. A considerable portion of the foot pavement in the town of Staines, in Middlesex, is also laid with concrete slabs, and forms a strong contrast to the old and dilapidated pavement of the flag system in the other parts of the town. There are numerous other examples which might be described, but we content ourselves with calling attention only to those few easily inspected concrete results of a by no means unimportant character.

Since the introduction of the "Patent Victoria Stone," now nearly eight years ago, something like a million of square feet has been laid in various places. To have accomplished as much in so comparatively short a time indicates that the material possesses merits of an unusually attractive kind, to have enabled it to prevail to such an extent against the various interests and prejudices with which it had to contend. A new material of this character, dependent in a great measure on the accurate and successful combination of chemical agents, and whose best effects can only be produced when the cementing agent is of unexceptionable quality, had, as a matter of course, its run of disappointment and failure. The experience, however, gained by the unavoidable vicissitudes attendant on a new process has developed its best and truest merits until the manufacture of this "Victoria stone" has become so accurate in character as to place its products in most favourable comparison with the best natural stones.

We shall shortly describe the process of manufacture, which from its simplicity will be easily understood. Before, however, entering on this description we should premise it

USES TO WHICH PORTLAND CEMENT IS APPLIED. 383

by saying, that the primary agent really is Portland cement, and its quality when used for this manufacture must be of an unexceptionably high character. Experience has shown this to be so essential to success that the most searching tests are applied before the cement is accepted for use. In fact, the cement must be of high specific gravity, *finely* ground, and capable of resisting the severest hydraulic or water tests. The possession of the most faultless cement, however, is not alone all that is required. The aggregate, or material of

FIG. 57.

WASHING THE GRANITE.

mixture, must not only possess a certain quality of hardness but its treatment previous to amalgamation with the, chemical agents is of a careful and painstaking character.

The material at present used is the Leicestershire granite, which is received from the quarries in the required sizes, somewhat resembling *pea gravel*. This undergoes a careful washing, so as to eliminate any clayey or other objectionable particles, and the process is performed as shown by Fig. 57. After this treatment the granite gravel, if it may be so called, is clean, somewhat gritty, and in a favourable condition for further treatment.

In the workshop, or moulding room, as exhibited in Fig. 58, it is mixed in varying quantities, according to the

Fig. 58.

FILLING THE MOULDS.

work required, with Portland cement, and moulded in wooden frames lined with metal. The slabs are so made that

USES TO WHICH PORTLAND CEMENT IS APPLIED. 385

both sides are alike, and when worn on one surface they can be turned upside down, securing, even in this apparently trivial arrangement, an advantage over the ordinary flag pavement, which has usually only one wearing surface. The slabs are kept for at least seven days before they are dipped in the silica bath or tanks, as shown by Fig. 59. In these baths they are generally allowed to remain for eight or ten

days, and when removed might be, if required, used in a month's time. There is much increased value, however, in allowing the chemical action to continue undisturbed for as long a period as possible. The manufacturers of this stone are fully alive to the advantages derived from keeping the

2 c

386 SCIENCE AND ART OF PORTLAND CEMENT.

Fig. 60.

slabs or other products for a considerable time after being finished, as the large stock exhibited by the sketch Fig. 60 indicates. Like good port wine they will improve in quality by keeping, and no one requiring and desirous of having a superior slab or other product of this process could reasonably object to pay the enhanced value its maturing must cost.

The materials required and now used for the production of the varied forms by the "Victoria stone" process are:

1st. Portland cement.
2nd. Leicestershire or Guernsey granite.
3rd. Natural Farnham silica.

The analysis of the chemical ingredients are:

No. 1. *Portland Cement.*

Lime	60·05
Magnesia	1·17
Alumina	10·84
Silica	24·31
Alkalies	1·54

No. 2. *Leicestershire Granite.*

We are unable to obtain an analysis of this granite, but the following results of experiments by Mr. Kirkaldy indicate that the granites now used, Markfield and Bardon, are of high compressive value.

"Results of experiments to ascertain the resistance to a thrusting strain of four 3-inch cubes of granite:

	Pressure on 1 sq. in. Crushed at	Pressure on 1 sq. ft. Crushed at
Bardon Hill	20,742 lbs.	1,334 tons.
Markfield	19,096 ,,	1,288 ,,
Mountsorrel *	17,533 ,,	1,128 ,,
Guernsey	15,002 ,,	970 ,,

* Average five 4-inch cubes.

"The specimens were cut and planed on all sides in a similar manner."

No. 3. *By Professor Attfield.*

Analysis of Farnham silica, two specimens:

	1.	2.
Moisture	26·00	30·00
Alumina, lime, magnesia, and iron	4·00	4·00
Silica	70·00	66·00

When dried each would contain 96 per cent. silica.

An examination of the sketches representing the various processes of this manufacture will show that no high technical skill is necessary to produce paving slabs, but, on the contrary, the whole operation is performed by ordinary labour unassisted by machinery of any kind. We should remark, however, that a steam engine is employed to grind and prepare in a mortar mill the crude silica as it is received from the quarries, so as to render its condition suitable for the bath. There has been no improvement attempted, owing in some measure to the opposition the introduction of an untried material of this character meets with, and also in consequence, probably, of the inventor believing that careless moulding was immaterial, as the silicating process converted the most carelessly moulded form into a compact petrified mass. Indeed, this confiding belief in the supposed all-powerful chemical agent encouraged carelessness in the selection and treatment of the other materials, so much so as to cause in the earlier days of the manufacture much loss and dissatisfaction. It was originally supposed that any binding agent, even lias lime, would, when submitted to the silicating process, produce slabs of a highly indurated character. These blunders, for they may be considered such from a practical point of view, ultimately led to a collapse of the company formed under Mr. Highton's auspices, and the process itself was saved from oblivion by the sale to the present company, who now carry on the manufacture at Stratford Bridge, in Essex; the past errors tending to stimulate them in their desire to produce goods of unexceptionable quality, and free from the defects inseparable from a too confiding reliance on the silica.

Some may regard the process of manufacture we have somewhat hurriedly described as too much dependent on manual labour, which might be profitably superseded by mechanical aid. In the comparative infancy of a manufacture of this kind it is not always advisable to introduce an expensive plant of machinery until a reputation has been established and tolerable certainty of success assured.

Paving, as practised in the metropolis and other large cities, is apparently a simple operation, and one which requires but little scientific skill for its successful performance. There are, however, difficulties of an almost insurmountable character to be overcome before a novel material, whatever its intrinsic merits, may or can be introduced. Vestrydom, or some other equally ancient fossilized body, has to be courted, coaxed, if not bullied, before a trial is permitted of even the best materials. A good example of the effects produced by this resistance to improvement is to be found in the state of the pavement extending from Ludgate Circus to Cannon Street station. Such a disgraceful piece of footway is not easily found, and those responsible for its condition should be punished for the many sprained ankles and other misfortunes of which it has doubtless been frequently the cause. But we forget we are in the "City," and should be prepared for such things, for we have just passed "St. Paul's," and feel ashamed at its external condition. Why cannot the clergy keep it clean? The revenues are sure to be well looked after and kept free of rust or any other degrading influence, but the noble pile itself is left to the mercy of a London atmosphere. Nature, with its cleansing rains and violent storms, does as much as it can, at all events enough to show the contrast between white and black. We might afford in these times of convulsive clericalism to change, or altogether dispense with, squabbling clergy, but to rebuild St. Paul's is now almost an impossibility. It would be a comparatively easy matter to restore the structure to its original colour and condition, and by the aid of the silicating or other cognate process maintain its integrity against the most violent and persistent attacks of impure atmospheric surroundings.

In our observations on this process we have thus far confined ourselves to the description of paving slabs, but it is not to be supposed that the capacity of the manufacture is limited to this useful production. Landings of considerable size are produced—steps, sinks, sills, and a variety of ornamental objects as well are readily moulded, but in a somewhat different manner. The more complex forms are made by using the mixture in another way. The finer portions of the granite are mixed with the cement in a much drier condition, in which state it is rammed into the mould by heavy and peculiarly formed broad-faced hammers. This percussive treatment has a most beneficial influence on the resulting products, and the sinks and sills obtained by such treatment compare favourably with those made from the best natural stones. The manufacturers appear ambitious to excel in the quality of their wares and with so wide a field in which to display their ingenuity we may look forward to many adaptations of the "Victoria stone," not only in a useful but ornamental direction. Feeling that progress is indispensable to success, they are aiming at colouring vases and other ornamental objects; and thus far their efforts have been attended with a fair measure of success.

The satisfactory results obtained through the agency of this and other processes elsewhere described indicate that great strides have been made of late years in the direction of artificial stone manufacture. The comparatively recent impetus given to this branch of constructive skill is due in a great measure to the improved knowledge of Portland cement and its valuable qualities. The stone maker is from increased experience better able to dispense with the unreliable cement of careless manufacturers, yet notwithstanding his utmost vigilance, is still occasionally put to great inconvenience and loss. He no longer reposes blind confidence in the maker's name, but trusts to his own knowledge as the best protection. Should this industry continue to increase in the same ratio that it has done during the last few years, the stone maker must make his own cement, when he could produce the required quality, and thus save himself much anxiety and loss.

It is but fair to the cement maker to mention that Portland cement best suited for artificial stone-making should possess properties that are not usually required of it in the ordinary concrete and similar constructive operations, and therefore, unless specially ordered, not as a rule usually supplied. The quality of stone-making cement should be of the most unexceptionable character, and the specification best suited for its obtainment ought to be as follows:

1st. Heavy cement of not less than 116 lbs. per imperial stroked bushel. This must not be obtained by any adulteration of slag or vitrified brick, but be pure cement produced from clinker out of the kiln.

2nd. A tensile capacity of not less than 300 lbs. per square inch after seven days' immersion in water.

3rd. Fineness of powder. Must not leave more than 10 per cent. of residuum in its passage through a sieve having 2500 meshes to the square inch.

In August 1869 Mr. Kirkaldy tested a specimen of the patent Victoria stone with the following results:

The surface exposed to the thrusting strain was 55·57 square inches.

Cracked slightly.	Cracked generally.	Crushed.
147·540 lbs.	178·620 lbs.	264·720 lbs.
2655 per sq. in.	3214 per sq. in.	4765 per sq. in.

Since that time much improvement has taken place in the quality, and no doubt more favourable results would now be forthcoming.

Many foreign works in paving have been executed, one of the most important being at Callao, in Peru, where Mr. Hodges (the constructor of the great harbour works there) laid down 15,000 concrete slabs 3 feet by 2 feet for paving the town. This pavement has given much satisfaction, and it wears well even in a Peruvian climate.

We have endeavoured to show in a somewhat cursory manner that hydraulic engineering has in the widest sense benefited by a free and confiding use of Portland cement. The dangers of faulty manufacture are encountered by the increased knowledge of the properties of the cement, and the command of the proper means of testing its quality. Confidence thus secured, combined with an awakening sense of the desirability of chemical knowledge, will ultimately, we feel assured, put beyond doubt or cavil the excellent properties of Portland cement and its various constructive combinations.

Although we are familiar with the great monolithic structures in Portland cement concrete erected around our coasts in harbours and docks, we fail to observe any progress in the direction of sanitary engineering, in the advancement of which we are all directly and personally concerned. The Metropolitan Board of Works it is true made a mighty stir and sounded a loud trumpet when they decided to use Portland cement mortar in the works of the outfall sewers. A grand opportunity was lost by the engineers of that body to display their appreciation of a comparatively new combination of valuable materials, at least new from our modern point of view, for concrete is nearly as old as civilization itself. The ancients bequeathed to us undeniable records of their confidence in this material, and used it fearlessly in aqueducts, cisterns, baths, and other sanitary structures. The still existing remains of these long-since constructed buildings — some even at a period anterior to recorded history—show us that for such application and purposes, concrete was regarded by them as pre-eminently useful.

The modern system of sewerage necessitates a most careful attention to the junctions of the subsidiary or house drains, to those huge gas-generating and rat-producing constructions called outfall or main sewers. It is not enough that these receptacles or conduits of filth should carry seaward to oblivion and utter annihilation the valuable means which should fertilize our exhausted soils; it is not enough that the ratepayers are saddled with the taxation of their cost, but in addition, which is more important and vital, they have their home atmosphere saturated with the poisonous gases of which these sewers are the permanently established generators.

In the registration division of London (according to the Registrar-General's return for 1876) there is embraced an area of 122 square miles, honeycombed by 2000 miles of sewers, over which traverse 1500 miles of streets, along which are built 417,767 houses inhabited by three millions and a half of human beings. Every square mile has therefore on the average 16 miles of sewers or something like $8\frac{1}{2}$ yards to every house. Within this district, so accurately defined and mapped, the population live at various levels, ranging from the minimum (at Plumstead) of 11 feet below high-water mark, to the maximum (at Hampstead) of 420 feet above the same datum.

In the old condition of sanitary science, and during what may be distinguished as the cesspool period, the junctions of the house drains with their points of deposit was a matter of comparative insignificance, for each dwelling, or at most a few associated together had to contend only with the difficulties of their own making. It is now quite otherwise, for every house and all which it contains is tied hand and foot in helpless bondage, against which its struggles for emancipation or remedy are helpless and futile, for dwelling and life are alike dedicated to the mighty Moloch of modern sanitary science. The main sewer is brought as near the populous centres as possible, and to these laboratories of disease and death you are bound to be connected. It is now no longer isolated and detached connection capable of control, but a huge combination of millions to assist in poisoning each other. The hideous "Frankenstein" has been invoked, and we can only now temporize with that which we have in our short-sightedness created. The periodical London floods of past times are not of the same character as the present, for their waters now contain a larger amount of liquid sewage produced by the omnipotent main sewer. In some of our streets during the hot season it is hardly safe, or at all events not pleasant to walk. Our windows must be closed, and the chance of purifying our homes during that season at least is impossible, for to be safely free from external and internal influences would involve practically an hermetically sealed dwelling, an obvious impossibility of course, but in the direction of which much improvement can be made.

Every one of us in London at least must be familiar with

the process by which, under the burden of high taxation, we are landed in this unfortunate position and rendered impotent in our own defence.

Under such circumstances as those we have described, the duty is imposed on each householder or his landlord of undertaking the compulsory process of effecting a junction with the main sewer. Such an operation should be performed with the greatest accuracy, and not only must the mechanical effluvia traps be perfect, but the junction drain itself unimpeachable in its perfectness. In a majority of cases, however, this is a mere fanciful estimate of what should be a stern practical fact. Houses hurriedly constructed on light, spongy soil, unprovided with stable foundations, readily settle, and in their gravitation dislocate the too flimsy junction pipes by whose agency all the sewage of every dwelling is disposed of. The whole machinery by which this most important duty is performed is faulty and dangerous. The hurried trench, almost always on made soil, is dug, and on it is placed a socket pipe, and the work again speedily covered over, these things being necessarily expeditious in character, for the pavement or street traffic must not be interrupted. Or the tenant is coming in at once, and the drain must be finished. The imperfection of such work is for a long time undiscovered, as any leakage is absorbed by the soil, and no injurious indications are apparent until after the absorptive properties of the contiguous ground are exhausted—until, in fact, the unavoidable congestion arises. The stoneware pipes are not accurate or reliable enough for so important a purpose. The process by which they are manufactured is inconsistent with true accuracy of form, and although some of the qualities of sewer pipes are creditable to those by whom they are made, it is impossible as a general rule to attain the necessary perfection in the sockets. When we consider the numerous joints required to accomplish the most ordinary house junction, any failure in their precision must result in the consequences we have endeavoured to describe. It is practically impossible to attain accuracy of form in the passage through the kiln of such goods as sewer pipes, for even in the more advanced stage of the potter's art, when the production of common terra-cotta is desired for architectural embellishments, great difficulty is encountered and the best selected results fail to accomplish what is much desired. Indeed it would almost require the painstaking industry of a Wedgwood to realize such an object. Other means by which we can reach the goal when honestly sought for is through the agency of Portland cement.

Germany, France, and countries in the New World have called in the aid of Portland cement concrete for sewers and drains. Through the agency of "Beton Agglomérés Système Coignet" many hundreds of miles of drains have been executed in France, and although that system cannot directly compete with our Portland cement concrete, owing to the greater cost incurred in its preparation, still it shows advantages over brick and pipe drains, not only in price, but quality. In the United States drain pipes have been made of cement concrete for some time, and so important and beneficial has this application proved that special machinery has been devised for their manufacture. In Brooklyn City something like two hundred miles of these concrete pipes have been laid, and with such success that it is regarded as the best-drained city in the United States of America. It must be mentioned, however, to show with what disadvantages this drainage has been effected, that Portland cement from England was used for the concrete, in conjunction with Rosendale cement, of much local celebrity. It is found that a judicious combination of these two cements results in a compound of much and unusual value.

The name of this cement originated from the fact of its

USES TO WHICH PORTLAND CEMENT IS APPLIED. 397

first discovery being made at Rosendale, Ulster County, New York. It is obtained from a water limestone deposited

Fig. 61.

in a series of beds highly contorted and complex. Like the lias formation of this country, it exhibits on analysis much difference of chemical value in the various beds, and their accurate and careful blending by the manufacturer is indispensable to the successful fabrication of a good cement.

An average of ten samples from separate beds of this cement or limestone gives on analysis as follows, the specific gravity being 2·8:

Silica, clay, and insoluble silicate	23·83
Alumina	4·53
Peroxide of iron	2·04
Carbonate of lime	36·75
Carbonate of magnesia	27·03
Sulphuric acid	0·81
Chloride of potassium and sodium	4·36
Water and loss	0·47

Extensive works, shown in Fig. 61, have been established on the Thames, at East Greenwich, by Messrs. Hodges and Butler, who manufacture drain pipes and other sanitary

Fig. 62.   Fig. 63.

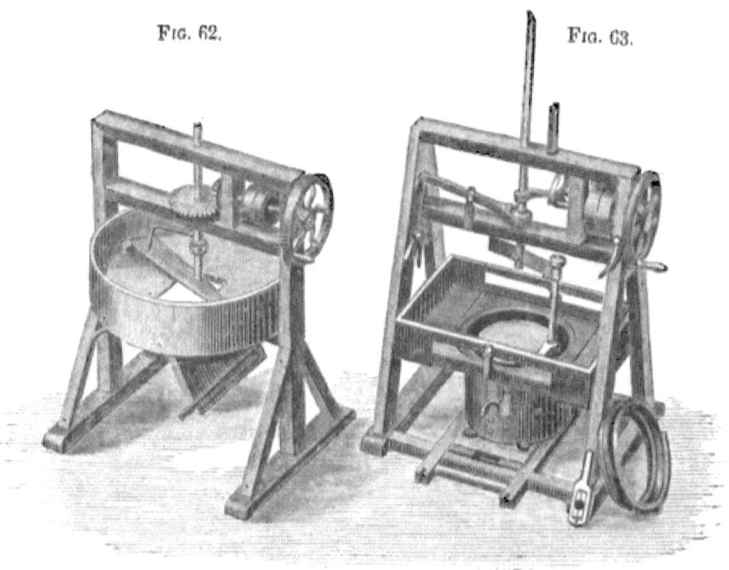

appliances by the American machinery, for which they hold the English patent. The produce of these works, more especially of the sewage pipes, is of the most satisfactory character. The pipes are socketed, but with flush joints, and as no disturbing influence arises after they are moulded, they continue as true in line as when they left the mould in which they were pressed together. The operation of moulding is performed in a careful and effectual way, and only the necessary moisture to ensure cohesion by the impinging rammers of the machine is used, so that no waste of the soluble portion of the cement is permitted or possible.

In Figs. 62 and 63 are represented the two machines used in this manufacture, by the aid of which the accurate pipes and other objects are made. Fig. 62 is the machine into which the necessary and proportional quantities of the several materials are put, which by its action are properly mixed. In preparing the flints or other suitable stones to mix with the cement, great care is bestowed on their reduction by the agency of a Blake's stone-breaker, and the fine dust incidental to the crushing operation is carefully excluded, as well as any clayey or loamy particles. The machine is driven by bevel gearing, which imparts to the spindle on which a knife or knives are fixed a rotatory motion, and when the materials are sufficiently amalgamated they are withdrawn through the lid at the bottom, as seen by the sketch.

Fig. 63 is the convertor of the previously mixed materials into the desired pipe, slab, or other form. Underneath the framework of the forming machine is the mould block, for making a pipe, and in the space between each the materials are carefully placed in the proper quantity, on which the rammer is made to act, rotating at steady and measured intervals until the space is filled, after which the top of the pipe receives by a collar the required split for junction.

Different adaptations of this simple mechanical arrangement enables the operator to make slabs of varied forms, as well as elbows, junctions, and other shapes of the most complicated character, as shown in Fig. 64.

400   SCIENCE AND ART OF PORTLAND CEMENT.

In due course the pipe so accurately put together is dipped in a solution of prepared silicate, thus hastening

Fig. 64.

the hardening process of the cement, besides increasing its ultimate strength.

The advantages which pipes thus carefully moulded

USES TO WHICH PORTLAND CEMENT IS APPLIED. 401

possess over the irregular and unsatisfactory stoneware pipe is very great. The flush and even joints, both inside and out, render any stoppage in the drain impossible. The

FIG. 65.

irregularity and unevenness of the stoneware or clay pipe is a fertile source of inconvenience and waste. It is absolutely impossible to render the joint of stoneware pipes either sound or capable of resisting the tendency to clog from pieces of timber getting into the crevices. In such cases the damage caused is incalculable. Neither are they water-tight under the best conditions of manufacture and laying. On the other hand, a silicated concrete, or stone pipe carefully made by these machines, is equal in its joints to the most carefully cast iron pipes. The sockets can be so accurately joined together that it is almost impossible to distinguish where the joint has been made. In experiments made to test the value and strength of the connection, it is found that the pipes will break more readily in their length between the joints than where the cement connects them together. Two pipes cemented together for some time, as shown in Fig. 65, have been submitted to a tensile strain, and their separation has not been effected even with a depended weight of a ton and a quarter. Their fitness to sustain the pressure of a head of water has been fully proved by an erection of thirty feet in height of pipes filled with water remaining perfectly tight, without the slightest indication of leakage or percolation, either at the joints or in the body of the pipes.

Another and most important property of the silicated pipe is its capacity to resist the action of gases and acids. In this direction there can be no question of its invulnerability to acid action, for in paper works Portland cement concrete is used in the construction of alkali vats, for which it has been substituted for York landing. Some years ago we submitted a slab of cement concrete made with gritstone to the influence of a chlorine bath for a fortnight, without its being damaged. It was desirable to obtain readily and at a cheap cost a large quantity of shelving, to be used in a new process at a Lancashire alkali works, as a substitute for Welsh slate, which at that time was in the required quantity unattainable. The author was consulted on the subject, and when the parties interested were about to proceed with the manufacture of concrete slabs, a new slate field was discovered from which the necessary quantity of shelving could be quickly obtained. The slate was at once capable of being used, and as time was an object, the intention of using concrete slabs was abandoned. We then felt, and our opinion remains to the present time unaltered, that but for the fact of the difficulty being overcome by the substitution of concrete for slate slabs, the cost of the latter would have been quite doubled. So much, then, for the protective influence of Portland cement.

Pipes made of Portland cement concrete will, when the conditions of manufacture are accurately performed, exceed in strength or capacity of resisting shock or pressure the best made stoneware pipes. For the purposes of extra junctions after being laid these pipes afford the most perfect facilities, as they can be bored without endangering the pipe itself, or disturbing its position in the smallest degree. Again, if damaged by any cause, they offer great facilities for reparation by the most ordinary class of unskilled labour; and, above all, when stoneware pipes, clay pipes, or vitrified pipes are subject to the deteriorating influences of sewage gases and other injurious action, tending ultimately to their disintegration and destruction, those made through the agency of Portland cement continue to indurate and improve in value. The maximum value of all these kiln-burnt wares is attained at the completion of their manufacture, and any injurious influences which they may encounter result in a depreciation of their normal or original value. Portland cement concrete, on the other hand, improves in strength, and we are yet ignorant of the period at which this improvement ceases. The process of induration is constantly in operation, and is influenced by the quality of the compound, and to some extent also by the circumstances by which it is surrounded. In sewer works or for drainage purposes the conditions for the improvement of concrete pipes or other of its structural combinations are favourable, for without the necessary moisture to perfect its ultimate crystallization, concrete cannot reach its highest point of value.

The numerous forms of bends and junctions can be readily moulded with the greatest accuracy, and in fact any desired shape of block or pipe is easily accomplished. The potter in dealing with the plastic materials of his process, assisted by the most perfect machinery, is unable to accomplish with anything like the same degree of facility or precision what the manipulator in concrete can without any effort effect. When the potter's task is completed, the produce of his handiwork is only entering on the dangers of its path. The crucial operations of drying and burning have yet to be performed, and all the accidents incidental to its manufacture and transit have to be encountered. The concrete pipes, on the other hand, may be made at any point where a sufficient quantity is required to warrant the establishment of a manufactory; and thus cost of carriage may be reduced to a minimum. On the other hand, stoneware pipes can only be produced in the districts where the proper quality of materials for their fabrication exist, or where at least they can be commanded at reasonable cost. These conditions, inseparable from such a manufacture, result in a localization of the trade, which in this country is limited to a few favourably circumstanced localities.

In the woodcut, Fig. 66, is exhibited a view of some more of the products from these works, in the shape of pipes, junctions, paving slabs, &c. Two 12-inch pipes support the weight of twenty 2-inch slabs, 3 feet × 2 feet, and the junction

USES TO WHICH PORTLAND CEMENT IS APPLIED. 405

pipes on the top thereof, or a weight equal to about 30 cwt., a pressure which they could not under any circumstance be

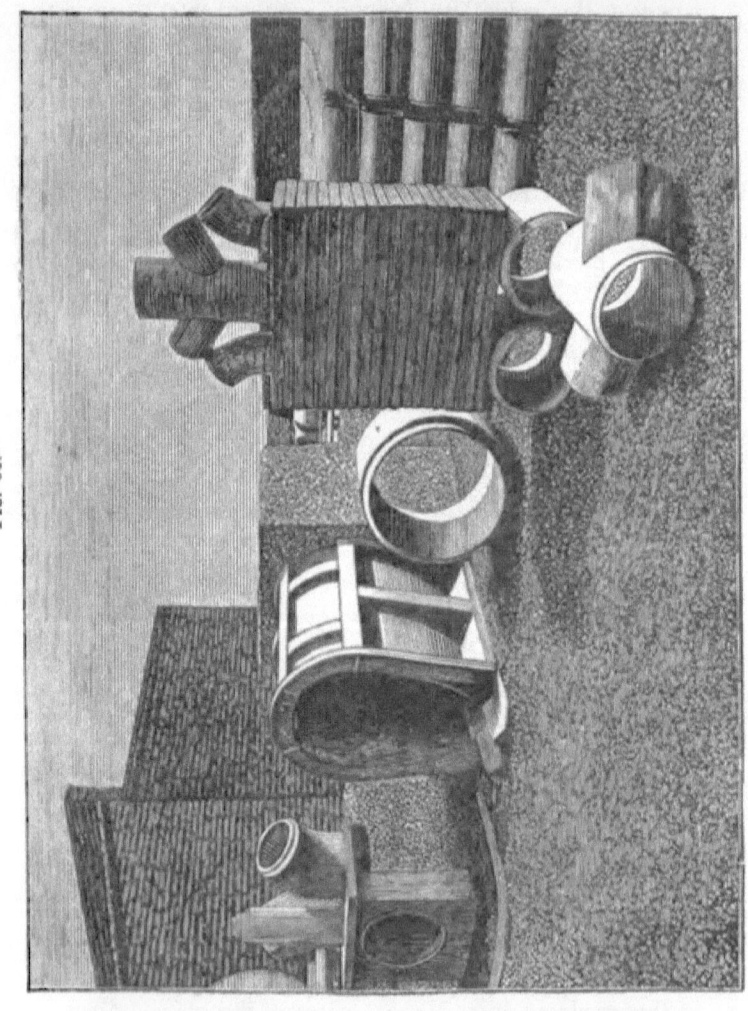

Fig. 66.

called upon to sustain in practice. On being first laid, even a few weeks after leaving the moulding machine, these pipes are competent to resist the above pressure, but afterwards they improve in hardness, and the limit or measure of this continuous induration is the final crystallization of the compacted mass. Accuracy of amalgamation of the materials, and their judicious selection, will, as a matter of course, influence the value of the products, for the machine can only ensure their true mechanical concretion.

An application of cement cylinders instead of iron for docks, wharf walls, and other similar purposes, has been executed with the most satisfactory results. Mr. Milroy, under the direction of Messrs. Bateman and Deas, has constructed brick cylinders, set in Portland cement, at Glasgow in a new wharf or quay wall of the harbour. The adoption of the brick cylinder in this instance was due to the circumstance of bricks being at that time unusually cheap, but an extension of the wall has been made in concrete cylinders, as bricks have doubled in price since the time of the first erection. Mr. Milroy, in conjunction with Mr. J. W. Butler, has for a long time advocated this system of cylindrical or columnar foundations, and the means and appliances they have designed for the successful execution of such works are most ingenious and satisfactory in character. Brick cylinders have been used for wells in India from a very remote period, and indeed in this country similar applications in a rough and careless manner have been made for well sinking, more especially in the rural districts of England. These works must, however, be regarded as very insignificant in comparison with the operations conducted by Mr. Milroy, whose words we will use to describe their nature and extent.

"The Plantation Quay is founded on a hundred brick cylinders, sunk in a continuous line close together, so as to form a length of 400 yards of quay. The wells are 12 feet in external diameter, and 2 feet 4 inches thick, thus having an internal diameter of 7 feet 4 inches. Their shape is circular, except at the points of contact, where they are formed with 'a tongue and a groove,' that is, with a square projection on one side, and a corresponding recess on the opposite side, alternately fitting into and sliding down the adjoining wells. The cylinders thus form a continuous wall, and the ground behind is protected from the disturbing action of the tides. This arrangement was suggested by Mr. Randolph, chairman of the Works Committee, in place of the wrought-iron tongue which was originally proposed."

In the further description of this work we will follow Mr. Milroy, with the explanation that the whole detail is equally applicable to the construction of the same form of cylinder or well in concrete.

Mr. Milroy, after describing the arrangement of platform, gantry, and steam appliances, proceeds as follows:

"On this platform the rings or annular sections of brick cylinder were moulded in frames of suitable size, constructed of wood, in four sections, bolted together. The woodwork consisted of two or three ribs, each of three thicknesses of 1½-inch planking, formed to the circle, nailed together and lined with 1-inch boarding. They were shaped with a recess on one side, and a projection on the other, to form the tongue and groove already spoken of. Annular layers of wood of the shape of the rings were fixed down to the platform, in order that their outer edges might keep the frames in place, while their inner edges served as guides in shaping the eye of the rings in building up the bricks. Four holes 4½ inches square were formed in the heart of each ring at equal intervals by means of mandrils set in sockets in the platform, for the purpose of cementing and jointing the rings together. When a frame had been fixed in place, bricklayers proceeded to build up the ring inside with bricks and Portland cement, using the cement freely so as to produce a smooth coating on the outer surface of the cylinder, to reduce the friction in sinking as much as possible. Each ring consumed nearly 2000 bricks: they were of the ordinary, instead of the radiated shape, which, had time permitted, would have been preferred as effecting a saving in cement. The cement, with a view to strength and quick setting, was mixed in the proportion of 1 cement to 1 of sand. When the brickwork had partially set, the screw bolts which secured the sections of the frame together were unfastened, and the sections removed. The rings were allowed to stand at least five days to consolidate thoroughly, and they could then be moved, either to be fixed in place or to be stored up for future use. It was 2 feet 6 inches high, and weighed between 9 and 10 tons. About 1200 rings, or about 3000 lineal feet of brick cylinders, were manufactured in this way, being about 8000 cubic yards, or about 11,000 tons of brickwork. Though only fourteen frames were used in moulding this mass, they kept their shape to the last, and were otherwise perfectly serviceable at the completion of the work. The removal of the rings was effected by means of the small traveller commanding the frame platform, which deposited them on a lorry; the lorry ran on rails up to the line of the cylinders, where the large traveller placed them in position."

What a wide field is opened up by this ingenious system of building, and it must eventually supersede many of the iron structures of which too free a use has been made of late by engineers. Architects seem to have run mad in the use of wrought iron in many erections for which it is inapplicable. The more durable cast iron seems now to be in the shade for structural purposes generally.

In the many applications of concrete for engineering and other works of construction, great advance has been made of late years, not only in the originality and boldness of the designs executed, but more particularly in the quality of the work itself. In house building, a more healthy system has become developed, and instead of a confiding belief in frames and other contrivances for the magic creation of all sorts and kinds of edifices, a more common-sense reliance has arisen for a good cement as the first condition of success. First secure that most necessary ingredient, and all succeeding endeavours will be safe and sure, for without it the most elaborate designs and cunningly devised machines will be worthless.

We do not purpose to discuss the house constructive aspect of the cement question, for sufficient information has during the last ten years been offered for the guidance of those desirous of building concrete houses. If we were to enumerate the various novel and recent applications of this material, the space at our command would be altogether insufficient. We cannot, however, refrain from alluding to a useful adaptation of concrete in the construction of a chimney shaft. It was erected at a cement works at Stuttgard, the height it was carried up being thirty-six metres. The chimney was required in the enlargement of the works, and not only was it built in cement, but all the foundations necessary for the new building were made of the same material until they reached the ground level. The progress made was one metre per day, and the smoke and heat from the various flues were passed through the chimney a few days after its completion. The mould used had the necessary tapering, and was externally of wood, the inside being sheet iron connected in a proper way by bolts and screws. It was one metre high and was moved each day. The cost was about 240l. At the junction with the flues, and until it reached the first set off, the interior was lined with fire-resisting bricks. A portion of Roman cement was included in the concrete mixtures to hasten its induration—a very doubtful proceeding from our point of view.

CHAPTER XXII.

Conclusion.

CONSERVATION OF LAND AND WATER.

THOSE outside of the constructive circle are now becoming familiarized with the various uses to which concrete is applied. In nearly every direction works of one kind or another are being constructed, and whether it be the visitor at his summer seaside resort, the inland tourist or the city inhabitant, all have opportunities of witnessing its various adaptations of sea-walls, docks, harbours, sewerage, aquariums, and dwellings. Our streets and pavements are no less matter for observation, and unfortunately in many instances the Londoner's patience is sorely tried by the continuous experiments in search of the still unsolved problem of the best material for paving. Notwithstanding this increase of concrete structures, much yet remains to be done; and when the still desired improvements in the manipulation of the materials are reached, the system must attain a position the extent of which we dare not venture to predict.

Our sea-girt isles, battered and bleached by violent gales from every quarter, suffer incessant degradation from the mechanical force of the storm-tossed sea, and if we were not familiar with the extent of damage caused by this unceasing agency, we have only to refer to historical evidence to ascertain the extent of land thus filched from our unguarded shores. Their future protection from this insidious destroyer is of easy accomplishment by the aid of concrete walling, which also could be applied more extensively to the reclamation of sea and esturial marshes.

Leaving our coasts, and proceeding inland, we find that the question of water storage remains in a grossly negligent condition. At the present time, when the water supply is, from its great importance, pressing to the front, and before its control becomes irrecoverably committed to the care of deep-well enthusiasts and their retinue of experts, the subject of our rainfall and its advantages should be freely and sensibly considered.

Nature's copious showers in our perhaps too highly favoured climate fall abundantly, and we negligently permit its waste by the absence of the necessary means for its conservation. Not only is the loss in a domestic direction immeasurable, but the waste of our land through this neglect is almost equal in extent to that occasioned by the sea. The rain water, in the hurried and unimpeded rush to its natural outlets of the ditch, stream, and river, carries with it the most fertile portion of our soils as well as a large amount of the costly manures of the farmer. This is a prodigal treatment not only of the natural agent, but the loss of the abstracted soil to which it imparts the moving impetus. This compound waste might easily be prevented, and enough water secured for domestic purposes, leaving a surplus for irrigating operations and fertilizing the soil, which its careful impounding would prevent from irredeemable loss.

A series of covered concrete reservoirs, communicating with each other by the aid of pipes or tubes of the same material, and judiciously contoured on the catch-water principle, would change the face of this country, besides affording relief to fever-stricken homes by a plentiful supply of water in their time of need. That the rainfall is more than ample for all moderate, and indeed extravagant wants, is beyond doubt; and its perfect utilization should no longer be delayed. It is needless in this place to discuss the amount of rainfall, but we may refer back to page 56 of this book, where an example of continuous downpour is recorded, showing that even in favoured Paris water storms are possible. Covered reservoirs are now regarded as indispensable for the economical storage of water, and at Montsouris, in the immediate neighbourhood of Paris, an extensive one has recently been erected, almost equalling in dimensions those of ancient Rome.

Water storage in any latitude, unprotected from the sun's rays, and free to be acted upon by the powerful natural force of evaporation, is a source of heedless and wasteful extravagance. The neglect of such protection involves an increase of capacity to meet the abstraction thus encouraged so as to provide for in dry seasons the necessary supply. An examination of reservoirs so inconsiderately constructed, proves their unfitness to contend with emergencies of protracted drought. Usually constructed at high elevations in mountain gorges, or at the mouths of ravines, they are costly in character, and under the most favourable conditions surrounded with ever-present danger. The state of these dammed-up artificial lakes during tempestuous weather is highly critical, but even that condition is less dangerous—at least to health—than in exceptionally dry seasons, when the reduced volume of water becomes simply filthy. It is at such periods too when the greatest amount goes off by evaporation, leaving the diminished contents further polluted by the impurities belonging to the pure water thus naturally abstracted.

The increasing scarcity of water in all directions, even from subterranean depths, should surely arouse the attention of the most apathetic to the coming danger, and encourage co-operation for general protection. It is only rain water after all; the nearest and the farthest and the deepest source alike owe their supplies to Nature's all-bountiful supply.

The "Fool's Paradise" of the measureless outpour from the "greensand," like the long-cherished inexhaustible "chalk" supply, are now things of the past. Stinted crops, unhealthy and disease-stricken homes, are the undoubted concomitants of drought; and the existence of both evils are due in a great measure to our continued apathy and indifference to the water-supply problem.

In many districts the form of reservoir might assume that of caverns hewn in the mountain rock or hill side, and lined with concrete where deemed necessary. The contents of these receptacles, conveyed through concrete pipes, would secure also an un-sunned supply of water not liable to increased temperature, owing to the non-conducting character of its conduits.

Street and Road Paving.

Of the many varied and changing systems of paving for our streets, nothing very definite has yet been reached, and we are still undecided as to the exact values of one or the other.

The existing plans now in the condition of test and trial may be thus classed:

>1st. Mineral.
>2nd. Oleaginous.
>3rd. Vegetable.
>4th. Metal.

The first, the well-known macadam system, and granite and basalt setts of various kinds.

Second. Asphalte of every quality, and various compounds of pitch and tar.

Third. Wood, generally firs of foreign growth.

Fourth. Iron, in tramways, curb-stones, and pavements.

With the exception of macadam, all of these systems depend more or less for their stability on Portland or other cement concrete.

Macadamizing is a species of concrete operation deficient only in the artificial binding agent, and its excessive wear and tear is due to the incessant disturbance and displacement of its component parts. Their constant action on and against each other produce, by attrition, a factor of disturbing and annoying character in the experience of us all. The "macadam" requires only therefore the mechanical and chemical aid of cement to perfect its character. The wear would then only be limited to the surface, and confined to the action produced by the traffic passing over it, and not increased by the wear and tear of its internal constituents.

The numerous qualities of "setts" obtained from various rocks, generally of plutonic origin, have had long and patient trials; but the desire to substitute for them asphalte and wood indicate that they are defective in character, and generally unsuitable for the traffic of great cities. Asphalte has had much attention bestowed on it, and in some directions its success has been tolerably satisfactory. Heavy and incessant traffic, however, develops its weakness and illustrates the disadvantages attending the use of an elastic covering under such circumstances. If it were possible to maintain its normal surface free from the disturbing influences of contraction and expansion, the wear and tear of such a material would be comparatively light. When, however, it is supported on a rock-like substratum of sound Portland cement concrete, its capacity of expansion is limited to a horizontal direction. The heavy traffic assists in this disturbing influence by flattening out, or, in other words, stretching the carpet to an extent beyond its inherent power of recovery. Hence the rolls or welts which ultimately crack and become developed into ruts. So long as the surface of this or any other kind of pavement can be maintained in a smooth and level condition, the possibility of degradation is, comparatively speaking, remote. Once break the skin of the surface, however, and immediately all the mechanical and other destructive agents attack it mercilessly, and its destruction is speedy and sure, as the condition of London streets too surely prove.

Wood paving is again coming to the front, and the only advantage it now possesses over its unsuccessful pioneer of former years is the concrete foundation. This is not, however, an unalloyed advantage, for it imparts to the pavement an amount of rigidity fatal to that vertical elasticity which is necessary to lessen the wear of the surface of the wood. Consequently the fibres of the cube are readily ruptured by the action of the horses' shoes and the passing carriages of all kinds. Like the joint of the slab or flag on the pavement, and the rut on the causeway, this is the forerunner of depression, increased as every wheel passes over it, until it becomes at last dangerous to use.

These various pavements are merely the precursors of what must eventually be the future pavement, namely, cubes of Portland cement, either by themselves or in conjunction with some of the above-named materials.

Manufacture of Cement for Special Works.

Portland cement has become so important a constructive material, and its cost now forms so large an item in engineers' estimates, that in many cases it would prove highly beneficial to make the cement on the site of the works or their immediate neighbourhood. In many cases this is not only possible, but, generally speaking, if the quantity of cement required is at all considerable, and especially in foreign works, it would result in much advantage. In one instance, in connection with the execution of extensive drainage works at Buenos Ayres, the author was entrusted with the preliminary examination of the local materials and the ultimate supply of the necessary machinery to the municipality of that city. The outfit for this experimental work consisted of—

    2 portable engines, 30 horse-power.
    3 pairs 4 feet 6 inch millstones and Hurst.
    1 brick-making machine.
    1 Blake's stone-breaker.
    1 Goodman's crusher and triturator.
    1 iron kiln.

A galvanized iron house, 100 feet × 30 feet, to hold the machine, with space sufficient to carry on operations under the dry process of manufacture.

The whole, after having been tested in England, and the men who were to accompany it trained to the work, was dispatched from Liverpool early in 1874. Owing, however, to the outbreak of the civil war in the Argentine Republic nearly twelve months were lost before a start could be made.

The whole was placed by the President under the management and control of Mr. W. F. Reid, who succeeded in making a good Portland cement, which was used in the sewer works. The raw materials were difficult to obtain, and costly, although there are many sources from which they might be obtained at more moderate rates.

The limestones and their analyses were as under:

No.	Carbonate of Lime	Carbonate of Magnesia	Alumina	Silica and Insoluble Silicates	Oxide of Iron and Alumina	Insoluble Residue	Moisture and Organic Matter	Loss	Locality	Analyst
1	79·92	3·66	..	..	..	10·64	5·78	..	San Salvador	Dr. Frankland
2	94·61	4·31	..	..	..	3·01	1·07	..	,,	J. J. Kyle
3	86·60	1·57	2·15	9·48	..	..	..	0·28	,,	,,
4	94·58	0·95	..	4·03	0·44	..	..	..	Maldonado	,,
5	49·60	0·39	48·05	..	0·85	..	..	0·11	Estanca Calera	,,
6	45·50	0·60	53·05	..	0·75	..	..	0·10	Punta Gorda	,,

SPECIAL CEMENT WORKS.

The analysis of the Toscas, from which the silica was originally intended to be obtained, were as follows:—

Nos. 1 and 2 analyzed by Dr. Frankland.
 „   3 „ 4   „   Mr. J. J. Kyle.
No. 5   „   Professor Rosero.

No.	Silica	Peroxide of Iron	Carbonate of Iron	Protoxide of Iron	Alumina	Lime	Carbonate of Lime	Carbonate of Magnesia	Protoxide of Manganese	Magnesia	Carbonic Acid	Potash	Soda	Organic Matter and Moisture
1	*Soluble in Acid.*													
1	1·36	2·94	..	..	2·67	10·24	..	..	..	0·83	..	..	..	..
	Insoluble in Acid.													
1	51·58	2·56	..	..	10·78	1·38	..	..	..	0·74	7·72	..	..	6·63
2	*Soluble in Acid.*													
2	0·97	2·96	..	..	4·61	9·07	..	..	..	0·72	..	..	..	..
	Insoluble in Acid.													
2	51·46	1·92	..	..	13·68	1·74	..	..	..	0·72	7·10	..	..	6·14
3	*Soluble in Acid.*													
3	..	..	3·48	..	3·65	..	45·75	1·80	..	..	..	1·57	..	..
	Insoluble in Acid.													
3	34·13	..	..	..	8·97	..	..	..	..	..	..	..	..	0·65
4	*Soluble in Acid.*													
4	..	..	6·30	..	6·00	..	19·25	1·05	..	..	..	1·26	..	..
	Insoluble in Acid.													
4	50·42	..	..	..	15·93	..	..	..	..	..	..	..	..	..
5	*Soluble in Acid.*													
5	5·07	..	..	2·35	1·43	..	58·46	..	0·06	0·96	..	0·13	0·49	..
	Insoluble in Acid.													
5	20·91	..	..	1·36	4·94	0·99	..	..	..	0·50	..	..	..	3·41

These Toscas occur in two conditions, one quality being nodular and hard, while the other is in a disintegrated state; Nos. 1, 3, and 5 samples being of the former quality, and Nos. 2 and 4 of the latter.

2 E

Sutro Tunnel Analyses.

In the beginning of the year 1874 the author was consulted by Mr. Sutro, the engineer and constructor of this great undertaking, as to the possibility of manufacturing Portland cement from materials in the State of Nevada. In consequence of the costliness of labour required to convert the timber necessary for the tunnel operations, it was considered advisable to substitute concrete blocks for lining and in the formation of the water channels, a most important part of the tunnel scheme. It was also intended to construct a concrete city at the mouth of the tunnel, to meet the wants of an increasing population, drawn together by the extensive mining operations of this highly favoured precious-metalled district. Congress had, in furtherance of this scheme, given a grant of land for the purpose of building the city, which, in honour of the great engineer, was to be called "Sutro City."

Numerous specimens of minerals were forwarded to England, and from which were selected the following samples, that, on being submitted to analyses, exhibited the following results:—

No.	Silica	Alumina	Peroxide of Iron	Lime	Carbonate of Lime	Carbonate of Magnesia	Sulphate of Lime	Magnesia with Silica	Organic Matter	Moisture	Combined Water	Remarks
1	78·6	7·1	2·5	2·1	..	..	..	..	..	4·3	5·3	White colour.
2	2·7	trace	trace	..	95·6	1·1	..	..	..	0·6	..	Limestone.
3	17·3	6·8	69·0	..	0·8	..	..	..	..	1·7	4·5	Ironstone.
4	9·9	1·7	0·7	..	13·0	0·9	59·7	..	1·0	..	13·1	Resembling fine chalk.
5	8·8	0·3	0·2	..	10·2	trace	63·8	..	0·2		16·5	Like chalk.
6	52·9	20·2	10·5	..	2·0	..	..	..	..	4·9	9·5	Crystalline.
7	75·6	3·0	3·7	..	1·0	..	0·6	..	0·3	11·4	4·4	White clay.
8	42·4	18·9	19·9	..	..	..	..	2·3	0·1	8·3	8·0	Red clay.
9	50·5	27·7	7·7	..	..	..	..	..	0·1	5·6	8·7	Blue clay.
10	91·3	5·2	1·4	..	..	..	..	..	0·5	0·3	1·3	Mine tailings.
11	78·2	5·0	4·6	..	3·4	0·4	0·2	..	0·4	2·9	4·8	Brick clay

A rich store, from which no difficulty need be found in making, not only Portland cement, but other more valuable products.

As an instance of the earnest energy of Mr. Sutro, it may be mentioned, that after the author's first interview with him, when informed what character of materials were required, in about three weeks' time they were placed in the author's hands: this celerity being due to the telegraph and "Pacific Parcel Express."

Kunkur.

An Indian carbonate of lime used largely in the works of military construction.

Its analysis is—

Carbonate of lime	37·01
Siliceous clay	49·80
Carbonate of magnesia	2·79
Oxide of iron and alumina	7·69
Organic matter	1·27
Moisture	1·24

The treatment and manipulation which this mild carbonate of lime undergoes at the hands of the native contractors appears to be destructive of its fitness for mortar. Disastrous consequences have resulted from its free use in the construction of barracks in India, attributed by some authorities to the existence of phosphoric acid. It is not necessary, however, to look about for a cause, for its analysis too clearly indicates its want of energy for mortar purposes. When Lord William Bentinck was Governor-General of India he used the kunkur in its natural state for road-making, and required it to be rammed or beaten, which resulted after rain in a hard cemented mass.

Artificial Stones better suited for warm Climates.

In Peru and other countries having similar climates, it is found by experience that Bramley fall and other English building-stones become degraded. Where cement slabs have been used in connection with works of construction in conjunction with English stones, they withstand the climatal influence uninjured.

Proportions of Raw Materials.

In the discussion of the various raw materials and their analysis we have not given accurate directions as to the proportions of raw mixtures, because it is impossible to give any precise formulæ where the value is so fluctuating and uncertain. Those who can ascertain the chemical value will have no difficulty in deciding on the correct proportions, but for the information of less favoured operators we give Mr. Lipowitz's rules, which will be found in a rough general way applicable for experimental purposes.

The carbonate of lime taken by him as an illustrative example was musselchalk of the Rhine, the analysis of which was—

Carbonate of lime	83·9
Silica	10·0
Alumina	1·5
Iron oxide	1·5
Magnesia	1·6
Water	1·5

The clay from Grenzhausen—

Carbonate of lime	0·61
Silica	68·20
Alumina	20·0
Potash	2·35
Iron oxides	1·78
Magnesia	0·25
Water	6·39

A brown iron stone from the Lahn—

Iron oxides	50·5
Water	9·0
Soda	10·0

The musselchalk contains, in round numbers, 84 per cent. of pure carbonate of lime. These represent:—

$$100 : 56 = 84 : x \;.\; x = \frac{56 \times 84}{100} = 47\cdot 04 \text{ burnt lime.}$$

If you require to make samples of cements of 55 per cent. chalk, then an addition to the 47 per cent. Ca. in the limestone would be necessary of $55 : 45 = x \;.\; x = \dfrac{45 \times 47}{55} = 38\cdot 45$, to make 85 parts of cement.

As the chalk already contains 13 per cent. of hydrates, namely, 10 per cent. silica, 1·5 per cent. alumina, and 1·5 per cent. oxides of iron, they must be subtracted from 38·45 per cent., or say, in round numbers, 39 per cent.; and if clay alone is added, 26 per cent. of it will be required. But as the clay contains 68·26 per cent. silica, 20 per cent. alumina, 1·78 per cent. oxides of iron, or about 90 per cent. hydrates, $9 : 100 = 26 \; x \;.\; x = \dfrac{26 \times 100}{90} = 28\cdot 88$ parts would have to be mixed with 100 parts of musselchalk.

Since the above was written, considerable increase in concrete works has been made in paving and sewering at Newport, Isle of Wight, and Bagshot in Surrey.

When railway officialism descends from its lofty pedestal and ceases to torment the public, we might perhaps see its attention turned to the condition of the permanent way of their lines. If concrete was used the dust would be diminished and the rolling stock preserved, as well as the comfort of the passengers increased.

INDEX.

ABERTHAW lime, 7, 12, 56.
Adcock's, Rowley, rag manufacture, 103.
Adie's testing machine, 279.
Albolith cement, 75, 166.
Alkali vats, 402.
—— waste, 120.
Alluvial clays, 85.
Alum waste, 120, 132.
Aluminate of lime, 151.
Amalgamation of raw materials, 144.
Analysis necessary to guide, 140.
—— Smeaton's method, 7.
—— desirable before beginning operations, 152.
Analysis of English and German cements, 252.
—— of chalks and clays, 254.
Anning, Mary, 35.
Antediluvian remains, 53.
Archer's stone-breaker, 196.
Arches, oven, description of, 237.
Arou, Jul. Dr., of Berlin, on clays, &c., 114, 148, 290.
Artificial stone preparation, 363.
Artificial stone suitable for hot climates, 419.
Ash mortar, 14.
Aspdin's patent and experiments, 14, 25, 31, 174.
Asphalte springs, 24.
Assyrian mortar, 24.
'Athenæum,' report of Faraday's lecture, 370.
Atlantic cable route, 49.
Atmospheric agencies, 46.
Author's briquette machine, 297.

BABYLON mortar, 24.

Backs or reservoirs, 229.
Bailey's testing machine, 286.
Barlow's, Professor, experiments, 334.
Barn floors, 128.
Baryta cement, 116, 363.
Basalts, 97, 100, 102.
Bath wells, 46.
Bazalgette, Sir J., 327.
Bed stone (fixing), 217.
Belidor, 2.
Beleek porcelain, 111.
Berlin testing machine, 291.
Berryton lime, 12.
Beryl, analysis, 85.
Bessemer iron process, 123.
Bischof, Professor, 53.
Black, Dr., 22.
—— Sea, 44.
Blake's stone-breaker, 179, 183, 186, 188.
Bleaching, Lancashire, 166.
Block beaches, 376.
Blue lias clays, 360.
Blue lias materials, 176.
Bock's kiln, 258.
Borne's kiln, 258.
Boulogne pebbles, 82.
—— cement works, 123.
Bramwell, Mr. W. F., 298.
Brandenburg clays, 91.
Brazilian pebbles, 85.
Breaking stones, theory of, 183.
Brick kiln, Rugby, 176.
Brick-making, ancient Egyptian, 174.
—— machinery, 174.
Bridstow limestone, 8.
Briquette press, 297.
Broadbent's stone-breaker, 192.
Brooklyn City, U.S.A., 396.

Buckwell's granitic breccia, 367.
Buckland, Dean, 370.
—— Frank, 37.
Buenos Ayres cement work, 415.
Builders' knowledge of materials, 350.
Building stones, 363.
Burr, origin of name, 213.
Burners, Thames and Medway, 249.
Bushel measure test, 297.
Butler and Hodge's works, 397.
——, J. W., and Milroy, 407.

CAEN stone, 34.
Calcareous streams, 43.
Callao Harbour and paving, 392.
Cambridge estuary, 87.
Cannon-street (London) paving, 389.
Carbonate of baryta, 114.
—— of lime, its extent, &c., 38, 41.
Carbonic acid, its properties, 244.
—— —— testing, 153, 308, 309.
Carboniferous limestone, cement from, 146.
Carboniferous limestones, 65.
—— caves, 69.
Careless use of cement, 350.
—— paving, 389.
Carelessness of tests by makers, 297.
Carrara marbles, 41.
Carr's disintegrator, 227.
Cast-iron, peculiarity of, 347.
Cast-iron, use of, neglected, 408.
Catseye, analysis, 85.
Cement, concrete, 409.
—— cylinders, 406.
—— increase of temperature, 309.
—— Portland, from blue lias, 168.
—— testing, 294.
Cesspool period, 394.
Chalcedony cement, 93.
'Challenger,' voyage of, 84, 108.
Chalk, 48, 52.
—— and clay process, 162, 181, 210.
—— and clays, analyses, 254.
—— its great value, 141.
—— marls, 80.
—— moisture in, 163.
Changing character of stone dressing, 216.

Chatham experiments, 14.
Chemical action of sea water, 377.
Chemistry, value of, to cement making, 361.
Chert, its characteristics, 146.
Chronological view of experiments, 29.
Clarke's, Dr., water purifying processes, 40.
Clay, its properties, 142, 152, 158.
—— moisture, 163.
Clee Hills, basalt of, 104.
Clerks of works, 367.
Clinker (mild), its importance, 324.
—— slippy, 342.
——, various qualities of, 198, 261, 345.
Coals and their varieties, 235.
Cockle shell lime, 6.
Coke and its quality, 236.
Coking ovens, 238.
Colson's experiments and paper, 311, 322, 340.
Comparative tests, 301.
Comparisons between wet and dry system, 174.
Concrete, improvement in, 320.
—— moulding, 384.
Confiding dependence on label, 357.
Conical mills, 226.
Conservation of land and water, 410.
Consumers' difficulties, 295.
Contractor's responsibility, 351.
Controllment of fuel, 249.
Coode, Sir John, 365.
Coral, Port Darwin, 51.
Cornwall clay, 88.
Cotswold Hills, 128.
Crossness reservoir experience, 359.
Crystals, their action, 347.
—— unchanged by washing, 212.

DANA, Professor, 33.
Danger of cement from sea action, 377.
Dangers of adopting novel forms, 305.
—— of mixtures in backs, 210.
Daniel, Professor, 73.
Danube River, 44.
Deacon's tests and experiments, 314, 317, 319.

Dead Sea asphaltum, 24.
Decantation process, 231.
Dee River, 105.
De la Beche, Sir H., 33.
Density of bricks, 268.
Density of granitic breccia, 377.
Derbyshire limestone, 132.
Description of millstones and gearing, 218.
Description of modern sewerage, 395.
Desideratum machines, 228.
Devonian period, 43.
Diabase cement, 113.
Diatomaceæ, 50.
Difference between wet and dry systems, 164.
Different results from various cements, 324.
Difficulty in changing established machines, 204.
Dinas fire clays, 89.
Dobb's patent, 15.
Dock works in concrete, 364.
Dome kilns, 243.
Dolomite, 112.
Dolorite, 112.
Dorking grey chalk lime, 12.
Dorsetshire clays, 88.
Dovehole's cement making, 179.
Dover Harbour, 365.
Double kiln process, 178.
Douglas Harbour (Isle of Man), 365.
Drainage works in concrete, 364.
Dressing millstones, 222, 226.
Drying ovens, 235.
Dry system of manufacture, 165.

EARTH, crust of the, 41.
Eathie (Scotch) lias, 59.
Eddystone lighthouse and rock, 1, 5, 361.
—— mortar tests, 5.
Edgbaston Vestry Hall, 103.
Egyptian pyramids, 24, 361.
—— or Nile Delta, 45.
Ehrenberg, the microscopist, 49.
Emerald, analysis, 85.
Erderienger's, Dr. L., experiments, 308.
Etna, Mount, 102.

Examination of clinker, 341.
—— of setting process, 347.
—— of mortar joints, 348.
—— of natural stones, 380.
Excelsior stone-breaker, 195.

FAIRBAIRN's experiments, 334.
Fairburn's steam brick press, 265.
Farnham silica, 374, 388.
Faroe Isles, lava from, 101.
Farraday, Professor, 370.
Fineness of cement, 322.
—— of raw materials, 176.
Finland granite, 111.
Fiureck's, Dr., experiments, 254.
Flints, 92.
Forming the raw materials, 167.
Frame building, 352.
Frankland, Dr., 417.
French method of making briquettes, 296.
French "burr stones," 213.
Fresenius's experiments, 39.
Frost's works at Northfleet, 14, 19.
Fruhling and Coy's testing machine, 284.
Fuel, 136.

GANGES River, Delta, 50.
Garnkirk clay, 91.
Gartcosh clay, 91.
Gas-lime waste, 133.
Gault clay, 90.
—— —— from wood, 133.
Geological observations on sea action, 376.
Geology, its importance, 38, 361.
German cement testing, 289.
Giant's Causeway, 41, 100, 141.
Gilby's, Dr., Analysis, 72.
Glat, River, 55.
Globigerina, 50.
Gniess, 119.
Goodman's crusher, 198, 200, 203, 205.
Gorham's process, 159, 211.
Granite, 109.
Granitic breccia stone, 371, 375.
Grant's experiments, 313, 327, 330, 335, 337.

INDEX.

Gray's stone breaker, 195.
Greenhithe cement works, 261.
Greenstone, 112.
Grinding machinery, its importance, 213.
—— (American) machine, 227.
—— retards in certain cases, 161.
Guildford chalk lime, 12.
Guthrie's brick press, 269.

HALL's stone breaker, 189.
Hanover cement experiments, 307.
Harbour works in concrete, 364.
"Hard work," 33.
Hartley, Sir Charles, 45.
Heat, 244.
Higgins's, Dr., experiments, 22.
Highly pressed bricks desirable, 265.
High Street, Bloomsbury, 381.
Highton's patent, 377.
Hodge's, Mr., 392.
Hoffman's kiln, 129, 241, 249, 254.
Hoogley, River, 50.
Hope's cubing jaw, 195.
Houses of Parliament, 73, 361, 364.
Hungarian cement, 83.
Hydrate of lime, 180.
Hydraulic engineering benefited, 392.
Hydraulic limes, 27.
Hydraulicity, how derived, 26.

IGNEOUS rocks, 147.
Impalpability of raw materials, 145.
Importance of fine cement, 183.
Improvement in cement, to whom due, 327.
Increase of concrete structures, 356.
Independent and novel tests, 331.
Indifference to the study of cement, 350.
Indurated chalk, 14.
Infusorial deposits, 81.
Injudiciously selected works, 139.
Innes's, Lieutenant, tests, 325.
Insignificance (comparative) of steam power, 46.
Instability of buildings from vibration, 219.
Intermediate system, 179.
Irish limestones, 39.

Iron pyrites, 143, 175.
Italian clay, 82.

JOHN, Dr., Berlin, 3.
Johnson's kiln, 233, 246, 260.
Joints of millstones, 215.

KENDAL slates, 105.
Kent's cavern, 69.
Kilned, double, process, 15.
Kilns, Rugby, 176.
—— Wakefield, 239.
—— Dome, utilizing heat, 240.
—— Hoffman's, 241.
—— Dome (ordinary), 243.
—— Johnson's, 246.
—— Lipowitz's, 254.
—— Bock's, 255.
Kirkaldy's tests, 387, 392.
Kunkur, 419.

LAVAS (various), 101.
Lancashire (Sutton) lime, 12.
Leger, M. D., 290.
Lias limestones, 57.
—— clays, 96.
—— deposits, their peculiarities, 53, 143.
—— shales, 170.
—— their suitability for cement, 168.
—— Leicestershire, 12, 95.
—— Lincolnshire, 12.
Liebeg, Professor, 33.
Liebeg's specific gravity test, 305.
Liffey River Works, 365.
Limestone and other sedimentary deposits, 147.
Limestones (sundry) analysis, 78.
Lime waste, 97, 120, 127, 180.
Lincolnshire estuary, 87.
Lipowitz's kiln, 254.
Liquid mortar, 11.
Liverpool corporation test, 316, 321.
Lewes clunch lime, 12.
Lewisham Terrace pavement, 368.
Lubricants for concrete moulds, 370.

MACADAMIZING roads, 104.
Macclesfield, Lord, 11.

2 F

426 INDEX.

Machine for making stone pipes, 398.
Machinery of reduction, 182.
Magnesian limestone, 71.
—— Scotch and Irish, 74.
Magnitude of Thames Cement Works, 174.
Main drainage tests, 174.
Manchester paving, 104.
Mann's gravimeter, 299.
Mann's, Mr., reward, 302.
—— gravimeter, accuracy doubted, 302.
—— experiments, 311, 323, 340.
Manufacture cannot keep pace with constantly changing tests, 331.
Manufacture from limestone, 178.
Manufacturing mysteries, 27.
Marsden's patent "Blake," 186.
Matlock baths, 46.
Mechanical reduction, 15.
Mediterranean Sea, 44, 377.
Medway clay, 19, 86.
Mersey clay, 90.
Metamorphic action, 40.
Meteoric agency, 46.
Methods (new-fangled), 159.
Metropolitan Board of Works' tests, 296, 322, 327.
Michaelis testimony to Smeaton, 2.
Micheles testing machine, 281.
Miller, Hugh, 34, 36, 59.
Millstones, 213.
Milroy and J. W. Butler, 407.
Mineralogy, 361.
Mississippi River, 44.
Missouri—singular occurrence, 62.
Mixing raw materials, 181.
Mode of ascertaining specific gravity of cement, 307.
Mode of moulding bricks, 178.
Modern system of sewage drains, 393.
Moisture (latent) in all materials, 164.
Mortar, its derivation, 24.
Mount Sorrell granite, 377.

NATURAL rock quality easily ascertained, 363.
Newcastle "Whin sill," 104.
Niagara, falls of, 46.

Nicholson's portable balance, 305.
Nile River, 45.
Northfleet Cement Works, 14.
North Welsh Cement Manufactory 160.

OBDURATE character of materials, 178.
Old red sandstone, 37.
Oolitic formation, 62, 64.
—— clays, 97.
—— limestones, 144.
Operative intelligence, 33.
Oven foundations, 237.

PALLANT's testing machine, 280.
Pasley's testimony to Smeaton, 3.
—— experiments, 87.
—— Sir C. W., 13.
—— slate experiments, 106.
—— testing, 21.
—— testing machine, 278.
—— water cement, 19.
Paris rainfall, 56.
Paul's, St. Cathedral, 6, 361, 390.
Paving stones, their lamination, 381.
Pengelly, Mr., 69.
Pharos, tower of, 9.
Phenician mortar, 24.
Piccadilly paving, 381.
Pit clays, 87.
Plaster of Paris experiments, 13.
Pomeranian clay, 91.
Porcelain clays, 88.
Porosity of concrete, 379.
Porphyries, 111.
Portable system of manufacture, 219.
Portland cement first used by Metropolitan Board of Works, 22.
—— benefited by chemistry, 362.
—— conditions of burning, 242.
—— from lime waste, 127.
—— its dispersion, 25.
—— its importance as an article of commerce, 29.
—— its increasing consumption, 28.
—— its nature and quality, 25.
—— its valuable properties, 27.
—— natural, 83.
Portland stone (merchantable), 9.

INDEX.   427

Potter's clay, 89.
Powder of cement, test by colour, 340.
Pretentious test by hand touch, 356.
Proportions of raw materials, 420.
Puzzolana terra, 9.
Pyrites iron, 58.
—— in lias clays, 96.
Pyrology, 85.

Quito mud vulcanoes, 55.

Ramsgate pier, 6.
Ransome stone, 370.
Raw materials—their characteristics, 31.
—— dangers in applying too much moisture, 100.
—— preliminary examination, 141.
—— question, 136.
—— seldom found together, 145.
—— stock of powder, 181.
—— their cost, 145.
—— their estimation, 148.
—— their varied treatment, 157.
Reckless use of Portland cement, 351.
Re-carbonating theory, 349.
Reducing machinery, 179.
Registrar-General's statistics, 393.
Reid, Major, 14.
—— W. F., 43, 416.
Reynolds, Professor, 282.
Rhine, River, 44.
Rhone, River, 44.
Rio Janeiro, Gneiss rocks, 42.
River action as a geological agent, 44.
Roman cement, 20.
Roman mortars, 23.
Roscoe, Professor, 417.
Rosendall cement, 398.
Ross, Major, 85.
Rowley ragstone analysis, 102.
Rugby Cement Works, 168.
—— cement, its position in the market, 172.
—— exploded system of manufacture, 170.
—— kilns and improved process, 171.
Runner stone fixing, 217.

Sanctuaries, St. Menan's, Cornwall, 112.
Sanitary science, ancient and modern, 392.
Santorin earths, 108.
Saxony clay, 91.
Scheibler's, Dr., carbonic acid tester, 153.
Schulatschenko's experiments, 42.
Schweitzer's analyses of chalk, 48.
Sedgwick, Professor, 104.
Selection of raw materials, 140.
Semple, 3.
Septeria, 98.
Serpentine, 114.
Shales, hard, 97, 180.
Sheppey, Isle of, stones, 86.
Sholl's direct-acting pneumatic stamper, 228.
Silica, 83.
—— Bath, its use, 379.
Silicated stone pipes, 296, 396, 403.
Silicating tanks, 385.
Simple brick-forming machine, 263.
Slabs (artificial), 403.
Slag, its utilization, 104.
Slag bricks, 125.
—— almost inexhaustible, 126.
Slags, Durham, 122.
—— English and Belgian, 123.
—— New York, 124.
—— Northumberland, 123.
—— South Welsh, 121.
—— sundry, 122.
—— Swedish, 123.
Sluggish draught of drying ovens, 176.
Smeaton, 1, 2, 9, 361.
Smith, Capt., translator of Vicat, 30.
Soap waste, 133.
Somersetshire lias, 12, 168.
Sorby, Mr., microscopist, 50.
Sostratus, architect of Pharos tower, 4.
Spectrum analysis, 41.
Spindle of millstones, 220.
Staffordshire clay, 89.
Staines' paving, 382.
Stephenson, George, 298.
Stettin cement works, 137.
—— Portland cement, 117.

Stettin testing machine, 288.
St. Leger's patent, 17.
Stone backing, 215.
Stone-breaking machines, 184.
Stoney, Mr., engineer, 365.
Stourbridge clay, 89.
Straub, A. W., and Co., grinding mill, 223.
Streets and road paving, 368.
Stuttgard cement works, 409.
St. Vicent's limestone, 12.
Suffolk bricks, 359.
Sulphur in coke, 237.
Sutro, Mr., tunnel analyses, 418.

TARRAS, 10.
Technical knowledge, 137.
Testing cement, 283.
Test mortar balls, 6.
Tests, necessity of uniformity, 328.
——, danger of departing from fixed tests, 331.
Thames, River, 44.
—— cement works, 137.
—— clay, 84.
Thomson, Sir Wyville, 108.
Thonindustrie Zeitung, report of, 302.
—— experiments, 302.
Thurston's, Professor, testing machine, 282.
Toadstone, Derbyshire, 98.
Tools (stone dressing), 222.
Toscas, 417.
Traventine, Italian, 46.
Tuscany, springs of, 46.
Tyrone slates, 106.

URICONIUM, 71.

Uses to which Portland cement is applied, 361.

VARIETIES of silicated stone pipes, 309.
Velocity of millstones, 216, 222.
Vesuvius, Mount, lavas, 107.
Vicat's experiments, 3, 26.
—— absolute strength test, 276.
—— needle test, 275.
Victoria stone, 377.
—— application unlimited, 390.
—— materials, 387.
—— process of manufacture, 384.
—— slabs and paving, 381.
—— stock, head, 386.
Virgil, 'The Georgics,' 128.
Vitrifaction of cement, 18.
Vitruvius, 3.
Vivaria, Island of, lava, 107.
Voelcker's, Dr., analyses, 64.

WAKEFIELD cement works, 31, 139.
Warwickshire cement works, 172.
Warwickshire lias, 93.
Wash mill, process limited, 177.
——, description of, 207.
Waste heaps, shales and clays, 129.
Watchet lime, 12.
Water from the London chalk, 46.
Way's, Professor, analysis, 48.
—— materials, 168.
Wellington, Duke of, 114.
Westminster Bridge (old), 10.
Whitehurst on toadstone, 99.
White's cement works, 19.
Wilkinson's experiments, 334.
Witherite, 114, 118.
Wood gas lime waste, 133.
Wren, Sir Christopher, 9, 361.